NELSON POINT

Portrait of a Northern Gold Rush Town

by

DAVID F. MATUSZAK

PACIFIC SUNSET PUBLISHING

FIRST EDITION

Published by
Pacific Sunset Publishing
RR 2 Box 668-A
Redlands, California 92373

Design by Words & Deeds, San Jose

All illustrations not otherwise noted are reproduced through the courtesy of the Plumas County Museum, Quincy, California.

ISBN 0-9633582-0-0
Library of Congress Catalog Card Number: 92-064297

Printed in the United States

10 9 8 7 6 5 4 3 2 1

In memory of my grandmother, Julia, who sparked an interest in my past.
And to my son, Christopher, who has given me a future.

CONTENTS

Appendices

LIST OF ILLUSTRATIONS.

PAGE.

AUTHOR'S NOTE.

I first panned for gold at the age of six near Junction City, Oregon. An old depression-era miner was my instructor in the art of fine gold recovery. He had supported his family through the Great Depression, averaging five dollars a day. Through my childhood and early adolescence I panned for fun during the warm Cascade summer afternoons. It was not until I reached my late twenties that I considered gold mining as a profitable adventure. I was one of the thousands of modern goldseekers during the gold rush of the early 1980's. The price of gold had skyrocketed to over $800 per ounce. I had been spending my summer vacations from the teaching profession working in the grocery store that had helped finance my college education. I speculated during the fall of 1980 that gold mining was my ticket out of the grocery business. It became just that.

I began to plan for the summer of '81. The most important decision to be made was whether to return to the goldstreams of Oregon where, during the summer of '80 while crossing the Cascades on horseback, I witnessed gold dredges strung out along the waters of my youth. Or should I pursue the fertile diggings of the original California gold rush? My decision was made in the spring while accompanying my state finalist to the high school state wrestling championships, which were held at Chico State University. While returning to the Sacramento Airport, I pulled off the highway at Butte Creek. My wrestler received a lesson in goldpanning. I discovered that the real gold was indeed in California.

My first summer in Plumas County was spent on the North Fork of the Feather River. I camped near Belden on Mosquito Creek. From there I prospected via gold rush era place names (e.g., Rich Bar, Yellow Creek, etc.). I knew nothing of their historical significance other than the intrinsic value of their names. Having experienced little success and finding most of the fertile diggings already claimed, by the following summer I found myself dredging the gravels of the Middle Fork of the Feather River near where the La Porte Road spans the river. This area had been set aside for "recreational mining," could not be claimed and, therefore, was open to struggling miners like myself. It became a twentieth century gold camp. Formerly canvas, tents were now made of nylon. Cabins had been replaced by campers, trailers, and motor homes. Keene dredges dotted the river so close together that you could hop from one to the other from Fells Flat to the confluence of Nelson Creek without ever wetting your toes.

My research for this book began in an attempt to locate historic diggings for further mining. I discovered that very little information was readily available about Nelson Creek. The mining reports from Nelson Creek further intrigued me. The creek was claimed solidly from the confluence of the Middle Fork to the confluence of Poorman's Creek. The miners from there with whom I had spoken were all very secretive and carried sidearms—indicating to me that their claims were valuable. The word on the river was that you stayed away from Nelson Creek unless you had business there. This mysterious behavior induced me to begin to dig for information that might help me to dig later for gold.

When the gold market crashed, a few claims on Nelson Creek were abandoned. I filed on a location there on July 4, 1985, which, I discovered years later, had ironically been the location of Independence Bar. I have since learned firsthand that those early mining reports had been greatly exaggerated and that gold fever and not gold had produced most of the enthusiasm. Notwithstanding those difficulties, the mining experience, together with camping at the site of the Pauly Hotel over the years, has fueled my interest in Nelson Point.

I focused my early research on the Plumas County Archives and the Bancroft Library of the University of California, Berkeley. I spent weeks at the Bancroft Library, reviewing *Hittell's Clippings, Hittell's Scraps, Bancroft's Scraps*, and *Bancroft's Notes.* Much of this material was uncatalogued and had been stored in the university's library warehouse in Richmond, California. Some materials took years to obtain. I also spent weeks in the Quincy public library where I completed a review of every microfilm record of the surviving county newspapers of the period. From these beginnings I pursued every available source of information regarding Nelson Point. Much of this research was accomplished during my annual summer sessions in the Plumas County Museum Archives. Finally, I am comfortable with the results. I have completed my research, knowing that inevitably more evidence will surface about Nelson Point. If it does, I would appreciate hearing about it.

PREFACE.

You won't find it on any contemporary maps. Not so much as a historical marker notes its location. All you will find are portions of a few building foundations as the only reminders that it ever existed. Yet Nelson Point was one of the most significant gold rush towns of the northern mother lode.

Early maps of California all suggest that Nelson Point was the most prominent mining location of the region. "A Complete Map of the Feather & Yuba Rivers," published in 1851, depicted "Nelson Cr." and Onion Valley as the only gold camps of the region. And Gibbs's "A New Map of California," printed in 1852, illustrated "Nelson's Cr." and "Poorman's Cr." as the only features in the vicinity. Regional maps continued to suggest that Nelson Point was prominent well into the early part of the century.

Yet, virtually nothing is known about this colorful gold rush town. The purpose of this work is to establish Nelson Point, where the evidence suggests that it belongs, as the single most prominent location in Plumas County during the early gold rush and to document its role in the development of that county. Through my portrait of Nelson Point as the quintessential northern mining town you will discover the gold rush experience.

Nelson Point was also known as Nelson Creek, Nelson's Creek, and Nelsonville. Early references to these names were made to the vicinity surrounding the confluence of Nelson Creek and the Middle Fork of the Feather River. Early residents considered the area to include not only the camps on Nelson Creek, but often those on the Middle Fork from Rich Bar to Poplar Bar and many of its tributaries such as Willow Creek as well. The town center existed in two locations, originally at the confluence which was often referred to as the "point." Later it was relocated to Lindsay Point on the La Porte Wagon Road.

I have made every attempt to let Nelson Point speak for itself. That is to say, that this is for the most part a compilation and distillation of primary source materials. In my research I tried to leave no stone unturned. I beg the reader's indulgence if at times it becomes difficult to discern precisely who is doing what and when. It is not for lack of effort on the author's part, but of the difficulties attendant to research based on partial and often inconsistent primary sources. I have endeavored to make things as complete as possible, while at the same time remaining faithful to to my goal of conservative historical interpretation.

Contemporary diaries, journals, letters and other accounts—particularly the Haun, Edwards, and Hambly diaries—along with government and other public statistics provide the foundations for this volume. Each individual wrote from a distinct perspective. Haun was the unrefined laborer turned entreprenuer; Edwards the socialite and political activist; Hambly the socially aware humanitarian turned spiritualist. Each was a miner. Each from different backgrounds. And each a Nelsonite.

I have used the following editorial standards throughout this work:

1. Excerpts from the James Haun diary appear exactly as they were written.
2. The Edwards, Hambly, and John Haun accounts appear just as they were written but for two minor alterations.
   The text has been justified. And because they did not regularly capitalized the beginning of a new sentence, I have interjected a double space between sentences in order to indicate where a sentence begins.
3. All other accounts are unaltered—original spellings are used throughout—except for text justification.

This book began as a short article and grew quickly into an obsession consuming a significant portion of my life in the past decade. It was a facinating investigation of a period and place that has grown close to my heart Occasionally, I became overwhelmed and discouraged about the progress of the project. One person more than any other encouraged me to press on. She is Linda Brennan, curator of the Plumas County Museum. My return each summer to Plumas County, was met by Linda's smiling face and soft spoken words of encouragement. This annual greeting kicked off each summer session of research in the archives. Her knowledge of local history and assistance in locating material was invaluable.

Next, special thanks is due to Scott Lawson, assistant curator of the Plumas County Museum. First, for the transcribed Haun diary without which the quality of this work would have been significantly diminished. Later, for the introduction to the Edwards and Hambly accounts. These discoveries have changed the very nature of my book in recent years and have resulted in many delays in its publication. Each of the past few summers while I prepared the final manuscript, he apologized for another related discovery including something new that he was certain would interest me. Each discovery was significant and induced me to rewrite the manuscript. Whether Scott is to be blamed or deserves credit for these delays will ultimately be determined by each reader. Scott, my sincere thanks. I have witnessed Scott develop into a first rate curator over the years His knowledge of the local history is second to none. And I am confident that we will all benefit further as he continues to make his own contributions to gold rush literature.

Sincere thanks are due to my friend and confidant, Don Chisholm. First, for his encouragement from the beginning of this project including access to Bancroft Library and for dragging me kicking and screaming into the computer generation. His technical advise was invaluable. Secondly, he edited the language of the final manuscript. I appreciate his regular encouragement to disregard the years I had already spent on this project and to complete it without compromising my own standards. For nearly twenty years Don has encouraged me to write. Any loose attachment to me of the label "academic" is largely to his credit. He has been my mentor.

There are many others who deserve to be acknowledged for their contributions to this book. They are: The Plumas County Museum Association for permission to reproduce the photographs appearing in this text, Bancroft librarians, Pat Van Deburg, Irene and Alex, Jim Lague of the Yuba Feather Historical Association, John Schmale, the Wells Fargo Museum, William Frank, assistant curator of western manuscripts at the Huntington Library, Ruth Haddick, Erle Pauly, and Barbara Pricer for her transcription of the John Haun Diary. Thanks to J.S. Holliday for *The World Rushed In*. He set the standard in gold rush literature to which all who follow him must strive.

It is impossible to individually thank all those who have contributed to this book over the past decade. The rest of them know who they are and to each of them my sincere gratitude. And finally my thanks, posthumously, to two men, Erwin Gudde for his *California Gold Camps* which served as a road map in the early stages of my research, and to Roy Post who died shortly after my interview with him, but whose pioneer spirit lives on in the pages of this book.

—all the appointments and appurtenances of a thriving and prosperous and promising young city,—and now nothing is left of it all but a lifeless, homeless solitude. The men are gone, the houses have vanished, even the name of the place is forgotten. In no other land in modern times, have towns so absolutely died and disappeared, as in the old mining regions of California.

—Mark Twain

CHAPTER I.

INTRODUCTION.

The California gold rush began with James Marshall's discovery at Sutter's Mill. Marshall had been hired by Sutter to construct a sawmill on the banks of the South Fork of the American River nearly fifty miles from Sutter's Fort. He discovered gold on January 28, 1848, during a routine inspection of his construction. Initially, the gold fever was subdued by uncertainty about the magnitude of the discovery. However, it soon became apparent that the find was indeed significant. The miners did not discover for many months how widespread were the deposits and, therefore, digging focused on the immediate vicinity of the original strike.

Sutter's closest associate, John Bidwell, speculated that the Feather River might contain the same hidden treasures as the American. He was right. In March he discovered gold at Hamilton Bend on the Middle Fork of the Feather River and soon relocated upriver to Bidwell's Bar where he made a rich discovery that would boom into one of the first gold rush camps. The rush quickly spread and within a few short years tens of thousands of goldseekers would establish camps throughout the western slopes of the Sierra Nevada.

Mines of the Feather River and its tributaries proved to be among the richest of the mother lode. Plumas County developed in the northern portion of the Feather River region. Gudde catalogued 188 principle mining locations in the county. Like other locations throughout the lode, some fizzled, while others developed into thriving camps. Those that lasted evolved into towns which, when the gold played out, either survived on other merits or were abandoned and left to fade into oblivion in the Sierra wilderness.

Nelson Point was one of those northern mining camps. Its evolution into one of the most prominent towns of the region and eventual decline offers a case study of the gold rush experience.

AN EARLY PLACE OF PROMINENCE.

Two noteworthy general descriptions provide the first brush strokes toward a detailed portrait of Nelson Point. Senator W.W. Kellogg, who settled in Plumas County at Rich Bar in the fall of 1858, remembered that:

> Nelson Point was originally located at the point of a hill, or bluff, at the junction of Nelson Creek and the Middle Fork of the Feather River, where excavations were made for the hotel and store site. The trail down the Hogsback from Onion Valley passed

> through the place. It was about as rough and inaccessible a place as could have been selected, but was a live business center for the miners of Nelson Creek, Independence Bar, Sherwin's Mill and for several miles down the Middle Fork; also for the Willow Creek, Bells Bar, and Poplar Bar mining population. Some of the placer mines in that region produced very large amounts of gold; in fact it is reported that more men made their "Home Stake" in that mining section than elsewhere.[1]

In his *History of California*, published in 1857, Hittell also described Nelson Point:

> Among the noted points on or near the Middle Fork were Stringtown about fifteen miles northeast of Oroville, Nelson's Point about twenty-five miles north of Downieville, and Gold Lake some twenty miles southeast of Nelson's Point...But the most noted locality, perhaps, on this Fork of the Feather was Nelson's Point, one of the roughest places in California. It consisted of a few houses, piled as it were without form or shape against one another on the sides of the precipices where the spurs of three steep mountains met at the junction of Nelson's creek. Its situation as a sort of center between rich localities made it well known; and as a subordinate place of supply for remoter settlements and especially as a place of carousal and wassail on Sundays for the miners of the rough and rocky regions round about, it became famous in the early times. There is no good reason to suppose that any one mining camp was much worse than many others, or that Nelson's Point ought to have had any special preeminence for vice or wickedness; but it was its fortune to be visited in its callow days by newspaper correspondents, suspiciously familiar with low vulgar haunts, who pronounced it "one of the most miserable cut-throat looking places within the range of the mining regions."[2]

In many respects Nelson Point represented gold camps throughout the mother lode. Yet, early the town established its prominence and a character all its own.

THE NATURAL SURROUNDINGS.

> *...a strange, white, porcelain-like rock, glinting like a tooth thrust through the red soil—was* quartz.
>
> —Bret Harte

Nelson Creek flows through one of the most rugged ravines in California. It is a wonder that a town ever existed in such an inaccessible location. Nevertheless, the ravine was overrun during the rush. In describing his new home, a miner named Lecouvreur wrote:

> I am pleased with the aspect of country. According to my estimate— uncertain at best, I know—our loghouse on Nelson Creek must be nearly on the same elevation with Grass Valley, and certainly considerably below "Onion Valley." The climate is mild and pleasant; the air is very pure; snow is to be seen only 400 or 500 feet above us on top of

the ravine where, however, it never entirely disappears; even not during the heat of summer. I am justified in calling this a ravine and not a valley, because the lower end is so narrow that the bed of the rushing stream takes up the whole width of it; the hills on both sides being more or less densely covered with firs.

There are numerous wild, charming spots on this creek above as well as below us, towards its mouth; it empties into the northeast fork of the Feather River, about eight miles from here. My time does not allow me to explore these places, and so I have to be satisfied when chance takes me to one of them.[3]

The earliest geological survey of the area was done by renowned state geologist, J. D. Whitney. He discovered:

Great masses of serpentine occur along the trail between Pilot Peak and Onion Valley; and between Nelson's Point a variety of magnesian rocks were noticed...

The cañon of the Middle Yuba (Feather ?)[4] is exceedingly deep, the difference of level between the river at Nelson's Point and the summit of Pilot Peak being fully 3, 650 feet... Nowhere in the Sierra do we find more stupendous examples of denudation than occur in the region north and northeast of Pilot Peak, in the cañons of the Middle Yuba

Frank Lecouvreur washing his clothes at his log cabin at Nelson Creek, 1852

Early photograph of Nelson Creek

> (Feather ?) and its branches. At Nelson's Point the slates stand nearly vertical, and crop out in grand masses along the sides of the cañon. But on the steep slopes on both sides the surface strata often curve, as if bent by sliding down the hill, so as to give the impression of a dip to the east, when in fact they stand perfectly vertical below. The elevation of Nelson's Point above the sea is 3,858 feet.[5]

Contemporary drawings of Nelson Point are rare.[6] However, James M. Thompson made a pencil sketch of Nelsonville while visiting his cousin, John Thompson. His sister, Mrs. C.M. Lakin, happened to be an art teacher. She developed the sketch into a large painting which now hangs in the Plumas County Museum. In the foreground appears the general store owned by Terwilliger and Thompson at the confluence of the creek and the Middle Fork. The scene is valuable for its illustration of the town, but it understates the true ruggedness of the landscape.

A more inaccessible ravine does not exist anywhere in the Sierra Nevada than that which was carved out by Nelson Creek. Yet, its ruggedness outlined its true charm. One can only marvel at the thought that such a community ever existed in this setting. Soon after its discovery hundreds of miners picked away at the banks of the creek and steep sides of the ravine in search of fortune. As the gold played out, the terrain ultimately determined the town's fate.

Lakin's painting of Nelson Point

NOTES.

1 Senator W.W. Kellogg, "Nelson Point," from an abbreviated version of an article read at NDGW (Plumas County Museum Archives: 1923), p. 1.

2 Theodore H. Hittell, *History of California*, 4 vols. (San Francisco: 1897), p. 102.

3 Frank Lecouvreur, *From East Prussia to the Golden Gate* (New York: 1906), pp. 241-2.

4 The Middle Fork of the Feather River.

5 Fariss and Smith, *History of Plumas, Lassen & Sierra Counties* (San Francisco: 1882), p. 134 (hereafter cited as Fariss and Smith, *History*). A 1936 U.S. Geological Survey marker now stands at a point near the Pauly Hotel site which determined the elevation to be 4068 ft.

6 In fact, this volume contains every known extant illustration of Nelson Point from the contemporary period.

CHAPTER II.

THE DISCOVERY OF NELSON POINT.

Prospectors first reached Nelson Creek in early 1850. However, neither of the two earliest recorded expeditions that traveled up the Middle Fork of the Feather River journeyed beyond the confluence of the creek. In February of 1850 Cotton's party set out from their camp on the Middle Fork:

> We traveled up the Mountain Ridge occassionally coming down to the stream to make explorations and went as far as Nelson's Creek, some ten or twelve miles above our Bar. At that time our camp was the highest up on the MiddleFork of the Feather River. We were out on this exploration some three or four days...On that trip, we sat down on the side of the mountain to rest and discovered gold in a crevice, and with a butcher knife dug out about four ounces. It seems strange, now, that this did not cause us to make a search to ascertain the scource from which the gold came down the mountain, and also to discover what accumulation of gold theremight be below on the bank of the stream, but our curiousity did not seem to be aroused by finding this gold and we simply passed on our way.[1]

Early in 1850, an Indian appeared in the Yuba region with gold nuggets from uncharted territory. This prompted a man named Marks to organize a secret prospecting party. Thirty men, one of whom was John Rose, who later related this story, departed from Marysville. Making their way to the region described by the Indian, they stumbled upon the mouth of Nelson Creek. However, nothing was found and they returned home, apparently never having discovered the rich placer deposits on Nelson Creek.[2]

Consequently, credit for the discovery of Nelson Creek was given to prospectors during the Gold Lake search which ocurred months later. When a man named Stoddard returned from the mountains with tales of a Gold Lake, miners throughout the Feather and Yuba regions abandoned their claims in search of "El Dorado." Hambly remembered the episode:

> ...I must say right here that there was a large influx of the Gold Lake Emigrants scattered in and around Plumas but I find such a class of Emigrants, or "wild goose chasers" are a poor class to settle down; one humbing begets another too often. Stoddard and his ignisfatuus is a myth, or analogy as held by some altho there are some that yet believe in the mythical Lake;...[3]

Though Gold Lake was never found, the search for it opened new areas to miners, including Nelson Creek.

Early in June of 1850, two men named Nelson and Batterton, and possibly one other, posted three claims on an "undiscovered" tributary of the Middle Fork of the Feather River. *The Charles Parke Journal* first recorded the event[4]:

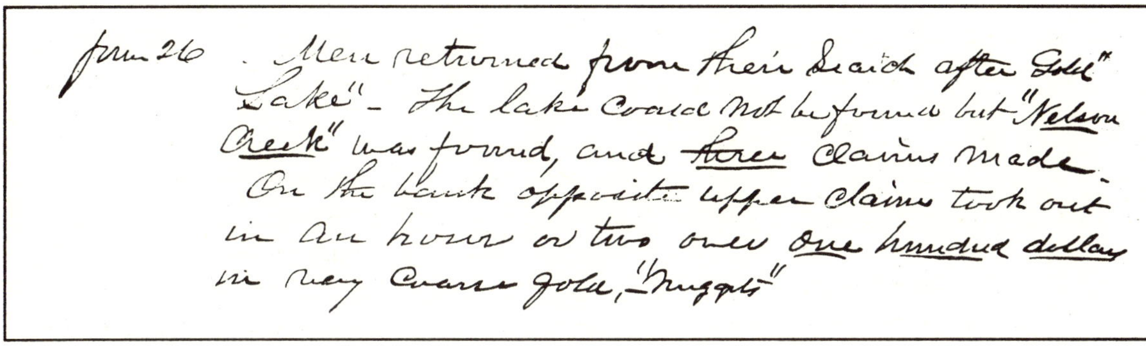
June 26 . Men returned from their Search after Gold" Lake" - The lake could not be found but "Nelson Creek" was found, and three claims made. On the bank opposite upper claims took out in an hour or two over One hundred dollars in very Coarse gold, "Nuggets"

Courtesy The Huntington Library

Nelson was followed by a flock of miners who quickly established Nelson Creek as one of the richest placers of the region.

The discovery of Nelson Creek came as a result of the search for the mythical Gold Lake. Miners from all over the northern mother lode abandoned their claims and rushed into uncharted regions of the Sierra Nevada. When it became clear that Gold Lake did not exist, hundreds of miners settled in to work the banks of the streams that they had dicovered along the way. Poorman's Creek, Hopkin's Creek, the upper extremities of the Middle Fork, Rich Bar and Nelson Creek became booming camps overnight.

NOTES.

1 Aylett R. Cotton, *Across the Plains to California in 1849 And After: An Autobiography* (Bancroft Library: unpublished 1906), p. 41.

2 Fariss and Smith, *History* (San Francisco: 1882), p. 149.

3 David William Hambly, *Hambly Autobiography* 1820-1890 (Bancroft Library: unpublished), 1853 (hereafter cited as Hambly, *Autobiography*).

4 Charles R. Parke, *Charles R. Parke Journal* (The Huntington Library: unpublished), p. 105.

CHAPTER III.

THE TRIP TO CALIFORNIA.

...I had only seven dollars in my pocket, five of which went to the purchase of a second-hand revolver, that I felt was necessary to signalize my abandonment of a peaceful vocation for one of greed and adventure.

—Bret Harte

Three routes from the east to California were most traveled by goldseekers: overland, the Isthmus of Panama, and around Cape Horn. Overland routes varied. Most ended in Sacramento. Those who came via the Isthmus of Panama first traveled by ship to that point. From there they traveled overland across the isthmus and caught a ship on the Pacific to San Francisco. The Cape Horn route was made entirely by ship around the cape to San Francisco. All three methods were long and tedious. Of the three, however, the overland route was certainly the most difficult. Of his travels, one Nelsonite wrote:

> You can form no idea of the labor, fatigue, trials and patience of an overland journey to this country. While traveling along the Platte for hundreds of miles, cold and rainy weather benumb your fingers while pitching tents, guarding cattle, preparing meals, gathering fuel so scantily distributed, and a thousand et ceteras blunt your facilities; and when the hour of quiet arrives at dark, you sink on your couch exhausted. It is the same when you reach the burning sand after passing the Platte; and, in addition to this, while traveling down the Humboldt (or Mary's River) the utmost vigilance is required to keep marauding bands of Indians from stealing or maiming your cattle; and you become wearied and worn out, so that if you lay over a day, you cannot collect sufficient energy scarcely to wash a shirt or mend your ragged and dilapidated garments. Any man who makes a trip by land to California deserves to find a fortune.[1]

Perhaps the best evidence that the overland route was the least desirable is that when the time came to return to the east few chose to return that way. By comparsion, nearly 90% returned via Panama.[2]

James H. Haun wrote a detailed diary in which he described his journey from Kentucky to Nelson Creek via Panama. He traveled up the San Wan River [sic] to Lake Nicaragua and on to Virgin Bay. From there he ventured by mule twelve miles to the town of "San Wan del Sea" on the Pacific. Haun boarded a ship there, cruised past Acupulco, across the Gulf of California, and anchored in San Diego in order to take on coal. From San Diego the ship navigated along the coast to San Francisco. At San Francisco, Haun boarded the steamer "Cumanche" for Marysville; however, it blew up ten miles short of its destination.[3] Marysville became the hub of the northern mining region. Once Haun reached there, he easily arranged for transportation to Nelson Creek.

And finally, William Edwards logged the mileage during his trip west from New York to San Francisco via the isthmus:

NY to Aspinwall	1978m.
Aspinwall to Panama	75m.
Panama to S.F.	3283m.
NY to SF	5336 [4]

Regardless of the route taken, the trip to California was prolonged and difficult. Were it not for the promise of fortune no sane man would have attempted it. Yet goldseekers came not only from the eastern United States, but from all over the world. Once in California the emigrant's travails had only begun. Hardships endured during the journey were but a prelude to what was ahead at the mines.

NOTES.

1 Alonzo Delano, *California Correspondence* (Sacramento: 1952), p. 22.

2 J.S. Holliday, *The World Rushed In* (New York: 1981), p. 530 (hereafter cited as Holliday, *World Rushed*).

3 James H. Haun, *James Haun Diary* (Plumas County Museum Archives: unpublished), March 23, 1853 through June 21, 1853 (hereafter cited as Haun, *Diary*).

4 William A.Z. Edwards, *William A. Z. Edwards Diaries and Account Books, 1843-1908* vols. 4-15 (Bancroft Library: unpublished), vol. 4, Feb. 18, 1853 (hereafter cited as Edwards, *Diary*).

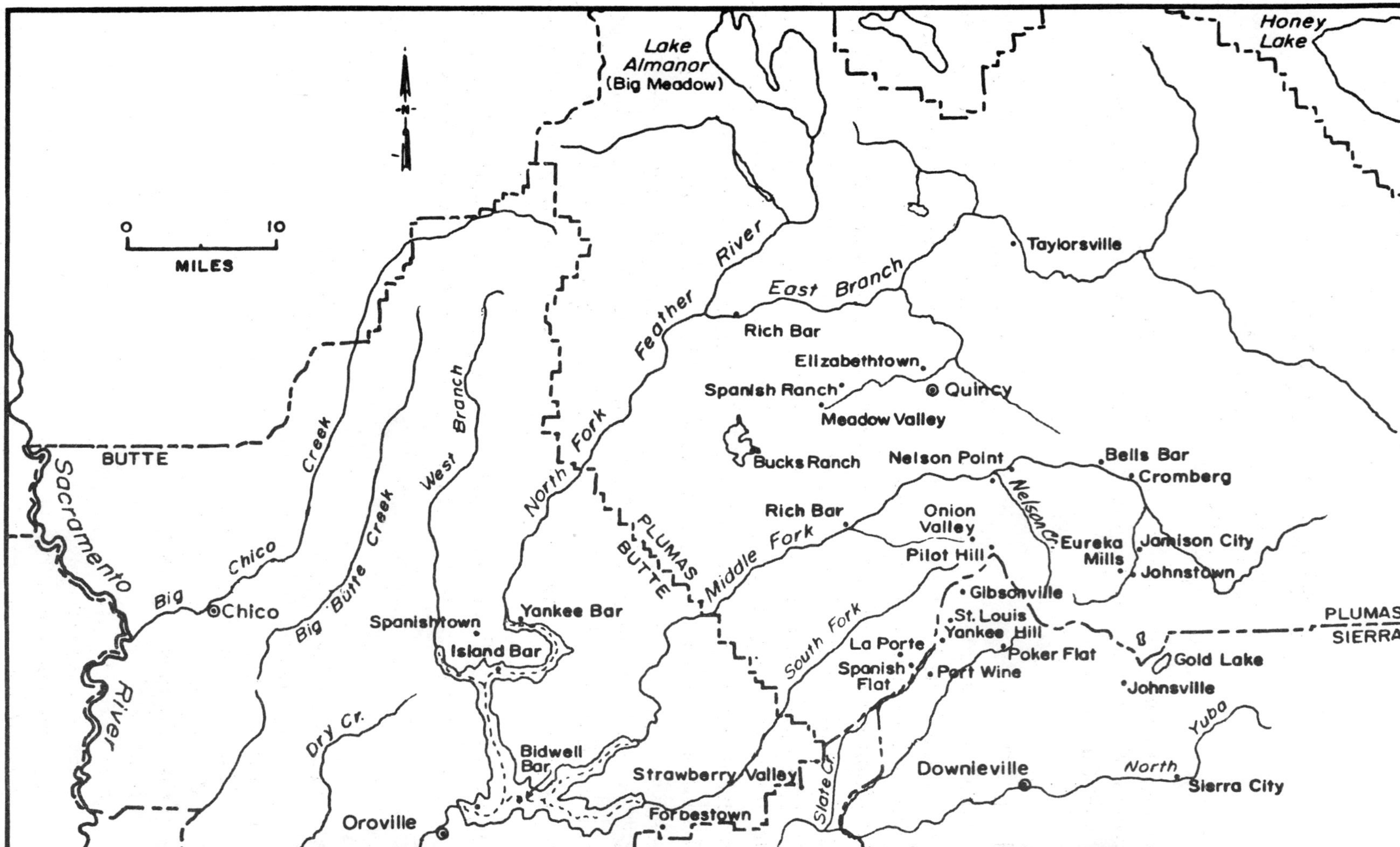

Adapted from E.G. Gudde

Plumas County and adjacent mining regions

CHAPTER IV.

TRANSPORTATION.

...by long mule teams that each team amounted to a procession, and it did seem, sometimes, that the grand combined procession of animals stretched unbroken from Virginia [City] to California.

—Mark Twain

The northern lode miner was constantly on the move. Whether traveling to his claim, his camp, to town, to the American Valley, to Marysville, or even to San Francisco, it is remarkable how often he traveled and how mobile he remained even under the most severe of terrain and travel conditions.

Plumas County exemplifies how transportation modes and routes developed throughout the northern lode. The first routes into Plumas County were mere foot trails. Next, passenger-trains and pack-trains of sometimes more than fifty mules wore in the trails until wagons could pass. Freight wagons and stages replaced the mule-trains; eventually private toll roads and public highways were cut in.[1]

By 1850, two significant trails had been established. The first path began in Marysville, passed through Strawberry Valley to Onion Valley, through Nelson Point and the Middle Fork of the Feather River, then on to the American Valley. The second route was Bidwell's Bar to Buck's Ranch, through Spanish Ranch, American and Indian Valleys, and on to the mines on the North and East Forks of the Feather River.[2]

Early routes to Nelson Point were well documented by Haun and Edwards, both of whom traveled them extensively. Haun says this of a trip from Marysville to Nelson Point in 1854:

Dec 54

We 6	Paid 10$ to ride up the Columbus house in the stage. Cold & foggy at marysville. Dinner 1$ at the oregon house. We got to Columbus at 8 o. Clk. P.M. took suppe 1$ & Bed 1/2$
Th 7	& breckfast 1$. I gave 64 to ride a mule to gibson vill & got my dinner there 1$ at 2 p.m. & then had to foot it 12

miles home whitch I did
a little after dark & carried
quite a pack. Spent 50 cts
for whiskey at the point.[3]

He later recounted the trip from the other direction:

Mar 55
Th 22 ...I was up before day & calld up all
hands in booth cabins eat breckfast
& I prepard to leave for Marysville.
I Shaw & John went down to the
point......I bid
John & Shaw good bye they start
ed home. I eat dinner with
Lewis cost 1$. With a heavy hart I
gazd at the high snow capd
mountain that lay before me
& then moovd steadly off to clim
over its height but before I had
near reachd its summit I dis
covered that I had lost my old bu
tcher knife out of the scabbard
a suffling through the brush
& snow as I had no father use
for the scabbard I took it off &
laid it in between two trees. I
made Gibson ville in good tim
though I would mire down
often in the snow. I stopd at
the mountain cottage 12 miles
from home. After dark I passd
a large house it seemd free to en
ter in I went & saw four mex
ican woman & a lot of dirty
miners dancing after the set
was out men & woman went
up to the bar & took somthing
to drink next follow a stag
dance that out I left & went
to bed...[4]

The following month, Edwards journeyed from Nelson Point to San Francisco, going via Onion Valley to Gibsonville where he took Spillman & Carr's Mule Train to

Rabbit Creek and then on to the Columbus House. There he hopped a stage on the road to the New York House, passed by the Keystone Ranch, on to the Peoria House, and stopped at the Western House in Marysville. This portion of the trip he made in two days. From Marysville he sailed by the steamboat Bragdon to San Francisco which lasted an additional day. Because Marysville and San Francisco were no more than a two or three day trip it is not surprising that visits to these cities, especially Marysville, were so common.

Local travel between Nelson Point and regional camps was made possible by a network of foot trails. Edwards noted the old trail from Nelson Point to American Valley over the mountain.[5] And Haun mentioned a route to Rich Bar which was directed over Slate Mountain.[6]

Local trails were managed by Nelson Creek's "pathmaster." J.L.C. Sherwin filled that position in 1855. Maintenance was financed by assessments of local residents. Edwards was notified by Sherwin to work on the road or be obliged to pay the latter $6.[7] In 1857 Edwards noted, "Silleger & I worked on the trail to pay our road tax..."[8]

FOOT TRAVEL.

Early travel on foot in the Sierra Nevada proved laborious. Trails were narrow, steep, and often life-threatening. Late in the summer of 1850 Kimball Webster departed from Nelson Creek with a companion:

> We started on the morning of the next day after the arrangements had been completed. His [pack] horse was at Onion Valley, about eight miles on our way. From Nelson's Creek up the mountain, about one and one-half miles, it was very steep. That brought us to the top of the mountain. From there we traveled to Onion Valley. There we took a little refreshments and then packed the horse and started for Grass Valley, about 18 miles distant...
>
> Soon after we left Onion Valley it commenced raining, which soon after turned into a wet snow. This melted nearly as fast as it fell, and being thinly clad as I was, I was soon wet to the skin and very cold and uncomfortable. I thought that if I could only get my blankets out of the pack to put over my shoulders it would be much more comfortable, and perhaps better for the health of a sick man.
>
> I spoke to Mr. Jewett about it and he replied that he couldn't get at the blankets without unpacking the horse, and that he thought I could get along without them. I said no more about the matter. We at length arrived at Grass Valley, thoroughly drenched and almost exhausted.[9]

During the rush, travel on Nelson Creek from claim to claim or to the "point" was by foot. Horses and mules were rarely kept for lack of pasture. Nonetheless, remarkable travel times were recorded on foot by miners such as Haun, Hambly, and

Edwards. They regularly walked, for example, from Independence Bar to Nelson Point, the way people today might walk to the corner market. Very often they packed provisions on their return:

> Sept 1853
> Sun 18 After breckfast we all went
> down to N. Creek to lay in a supply
> of provisions. The mountains are verry
> steep & high yet we pack 50lbs very easy
> on our backs...[10]

One explanation for the short travel times between locations along the creek was the use of wooden water flumes.[11] The miners regularly walked along or when dry inside of them— particularly Sherwin's Flume.

PACK-TRAINS.

Pack-trains, ranging in size from two to more than fifty mules, transported goods to the mines. Three hundred pounds was considered a good load for each mule, but occasionally a piano, printing press, or a safe was tied to a mule.[12] The California State Agriculture Society's 1858 Visiting Committee made this observation at J.W. Thompson's farm, also known as the Illinois Ranch:

> He keeps a train of pack-mules, now numbering forty, with which he freights goods from Marysville into the Valley.
>
> Although it is now in June, the season will not be sufficiently advanced for him to commence running under ten days or two weeks. A train of pack-mules is a curiosity in its own way. Their saddles are composed of a pair of leather-sacks, about two feet square each, stuffed with straw or some similar substance, till they assume the form of a common pillow, with feathers nearly all shaken towards the ends, and, when on the mule, resemble such a pillow, hanging across the arm. When the train halts for the night, and is unpacked, the saddles are taken off and set upon the ground in a row, in the same position as when upon the animal. When ready to pack up, the muleteer utters a sort of cry, groan, or some other indescribable note of command, when the mules, being released from their pickets, will go and place their heads over their saddles, standing in a row abreast, nor will there be any peace till the last one is in his place. There is no direction given, or pains taken, but every one knows his own place in the train, and his own saddle, and will be content with no other.
>
> These mule-trains are indispensable to the miners. Without them they could never carry provisions into those terrible fastnesses so as to sustain life while extracting the gold.
>
> Illustrative of this, in the winter of one thousand eight hundred and fifty-two and one thousand eight hundred and fifty-three there being a scarcity of provisions, Mr T. took his pack-train and made his way out of the mountains to Marysville, loaded them

with provisions, and started back. Meanwhile a very heavy storm had resulted in a great fall of snow in the mountains. After several days of toil they reached the top of the last great divide, but to their utter dismay they found that the wind had blown the snow from the side they had just climbed, over the top, and lodged it in a drift which broke off more than forty feet perpendicular. A long time was spent by the men in endeavoring to get such a trail as the mules could be induced to follow down. But in vain—none of the usual appliances could induce a single mule to make the attempt. The case to be decided was, whether the train should be turned back and leave large numbers of persons, (depending for life upon its speedy arrival,) to starve, or resort to some new measure. At length the order was given: "Overhaul every pack, and make it as tight and fast as possible," and as soon as this was done: "Blindfold every mule," which being accomplished, one was led to the brink, and stopped, with the right side to the precipice; three men were placed at his left side, and the order given; "Push him off," and away he went, rolling, first over his pack, and then his pack over him, till he reached a stopping-place, fifty feet below; then another and another, until the whole forty were thus tumbled into the valley of death with the means to preserve life. But the strangest part is yet to be told. Such is the skill attained by experienced packers, and such the hardiness of this invaluable animal, that this whole train was thus precipitated down the immense gulf of snow without breaking a single pack, or seriously injuring a single animal, although many of the loads weighed more than two hundred pounds each.[13]

While waiting for traveling companions to rendezvous with him at the Wisconsin House in 1852, Frank Lecouvreur watched pack-trains pass on the road before him:

...Away frrom the road no sign of life; on the road itself, however, it was different. Trains of pack-mules would pass us every now and then, led by "muleros" (mule drivers) in gay old Spanish costumes, their dark, sunburnt faces shaded by the broad brimmed felt hats, the long rifle in front on the saddle, and knife and pistol in their belt. They reminded me of Italian bandits as they galloped past us on their small, half-tamed horses, now in front and behind the long line of heavily burdened mules; now keeping them back and now urging them on by the use of their lassoes, their most dangerous weapons, which every one of them carries on the pommel of his saddle. They are a wild, dangerous set, these "muleros." Mostly Mexicans, they have very little love for the foreign intruder in—what they still consider—their own country; and when a man is alone on the road, he does well to keep out of their way—for they are exceedingly handy with their long knives, and a murder will not weigh heavily on their consciences.[14]

Hambly noted the reliance on pack-trains for supply of provisions to Nelson Point:

On the 17th of April [1854] Fred. Mandonald,s train came down the hill it was welcomed by all hands on the creek—provisions, was not very plenty—but prices were up.[15]

The following spring Edwards recounted the arrival of the year's first train:

> Joe Davis arrived at Independence Bar with his mule train 6 days from Marysville the first train for the season and 27 days earlier than last year[16]

The rugged terrain and narrow trails often caused the animals to fall off the trails and tumble down the mountainsides. Edwards witnessed one such episode:

> ...Saw two trains of Mules come in one mule rolled down the hill from the trail and lodged between Lewis & Ellis's store and the bank I assisted in unpacking it after which a part of the side of the store was removed and thus mula was extricated another rolled down from the trail & smashed the Sherwin flume and lodged therein...[17]

In later years pack-trains traveled from eastern routes as well. Pack-trains operated by G.E. Smith out ofVirginia City supplied Nelson Point in 1862 by way of La Porte.[18]

MOUNTED EXPRESS.

In the early days, mounted express agents were common throughout the northern lode. Initially they traveled at a leisurely pace—perhaps two trips per month. However, competition between expressmen ultimately resulted in faster and more frequent trips. "Letters, newspapers, small parcels, and gold-dust were the articles carried by the expressmen, the postal business being the most important and the most remunerative."[19]

Nelson Point boasted of its own express:

> "Miners Express" Nelson Creekers have established a mutual express company—one of them comes to Quincy each week for the mail.[20]

The most prominent expressman of the region was Frank D. Everts, who made his headquarters at Central House ranch, located twelve miles above Marysville. Precisely when he began operations is unclear, but it is known that Everts & Co.'s Express was in full operation by August of 1851. In 1853, when Nelson Creek and other nearby camps were booming, Everts required a partner to meet the growing demand. Thus, Everts, Snell, & Co. was formed. They ran from Marysville to Onion Valley and Nelson Point. Later, Everts became an agent at Nelson Point for Adams & Co. Initially, Adams & Co. did a lucrative business in banking which included purchasing gold-dust. However, the firm failed in 1855, causing a financial panic in California, ruining hundreds. Before closing their doors, they instructed Everts to forward all money and dust to the central office. Foreseeing that it would all be absorbed by the company, Everts notified all his customers of the situation and permitted them to withdraw their deposits, thereby saving the holdings of scores of hard-working miners and securing Everts a place in the miners' hearts.[21]

Haun had utilized Everts' services and recorded the Adams & Co. failure:

Dec 54
Sun 10 ...I wrote a letter to H.P. and
sent him a check for 1500$
by F.D. Everetts & Co express
to be loand at 3 per cent a month.
Feb 55
Sat 24 ...I was down to the
point & larnd for the first tome
Adams & Co Banking Express
has faild...[22]

Edwards recounted the rumors generated by the panic:

> Ellis arrived and announced the failure of Adams & Co Wells Fargo & Co & Page Bacon & Cos...The announcement of the failure of different banking & express establishments created quite an exitement and the greatest scraching & grabbing after gold dust or money in any shape ensued.[23]

In 1856, an express was run between Nelson Point and Gibsonville by Bill McCartney and Jim Grace.[24] The extensive number of noted expressmen in the region clearly shows that not only was demand for such services high, but the business was extremely competitive:

Frank and Henry Everts—Everts and Company's Express—1850-52: Marysville to Feather River Camps (Gibsonville, Hopkins Creek, Onion Valley, Port Wine, Nelson's Creek, Poormans Creek, Pine Grove, Strawberry Valley, Spanish Flat, St. Louis.)
Everts, Snell & Company's Express—1853-54: Marysville to Feather River Camps
Everts, Davis & Company's Express—1855: Marysville to Feather River Camps
Everts, Hannon, Wilson, & Company's Express—1856: Marysville to Feather River Camps
Everts, Wilson & Company's Express—1856-62: Marysville to Feather River Camps
Fargo & Companies Feather River Express—1853: Marysville to Feather River Camps
Copely's Miners Express—1855-6: Marysville to Feather River
Muma & Company's Express—1856-58: La Porte to Nelson Point, etc.
Whiting & Company's Feather River Express (Fenton B. Whiting)—1855-68: Connected with Everts and Company at Quincy
Holland, Morley & Company's Express—1862-63: Marysville to Feather River
Holland, Wheeler & Company's Express—1863: Marysville to Feather River
Wheeler's Express—1863-65: Marysville to Feather River

Wheeler, Rutherford & Company's Express—1864: Marysville to Feather River
Knowles Snow Shoe Express—1869-72: La Porte to Quincy
Brewster's Saddle Train—1858-67
Brewster's stages—1867-71

However, through a process of consolidation Wells, Fargo & Co. controlled the express business of the region by 1868. At the same time, a gradual transition from mounted express to the use of freight wagon occurred.[25]

FOUR-LEGGED TRAVEL.

Horses and mules contributed a significant portion of the transportation in the region. Records document that Thompson operated stables at Nelson Point and the Illinois Ranch.[26] Because Thompson also ran a saddle-train to La Porte it is likely that he also operated a stable there. Accounts indicate that he may have conducted a similar operation at the American Ranch. From these stables Thompson rented horses and mules.

Edwards documented his own travel by quadruped:

> Rode leisurely to Nelson Point [from Illinois Ranch] paid for horse hire $7.50...[27]
> ...E.A. White & myself procured mules at Nelson Point & proceeded to American Valley discharged our mules at the American Ranch. Mule Hire 3.00[28]

Travel by horse or mule along Nelson Creek was not common because the terrain was so irregular and steep, but occasionally it did occur:

Feb 56
Th 28 ...Fox came up in the P.
M. to get us all to go down to the
point to a dance, we refused to go
but him & Frank came back wi
th mules for them to ride, so
we all fixed up except Dave &
went down...
Fr 29 I came home about noon. Soon after John
& Carroll came over from Massac
& left Bray there. Wife came ho
me on a mule...
May 57
Sat 16 ...we walked up
to our diggins, saddld up the
Black horse & home we went
Su 17 I, John Duesler & Hunti
ngton went up to see the

new diggins on horses, we
we came back middle p.m.
took dinner, Lem went
with John part of the way
home to Nelson on horses—[29]

PASSENGER (SADDLE) TRAINS.

Passenger travel was initially accomplished by means of the passenger-train—sometimes referred to as a "saddle train." Modelled after pack-trains, saddles replaced pack frames on mules. W.S. Dean operated one of the earliest such enterprises in the region. In 1855, he began a passenger saddle train between Bidwell Bar and Quincy. By 1856 the path had widened enough to allow Dean to switch to stages.

However, because the terrain was significantly more treacherous, saddle trains could not widen the trail to Nelson Point (from Rabbit Creek to the American Valley[30]) sufficiently to allow for wagons. Consequently, the trail leading directly to Nelson Point never progressed beyond the use of passenger-trains. Although saddle trains were well established on this trail by the early 1850's, an indirect consequence of the physical constraints was the relocation of Nelson Creek to Lindsay Point.

Haun recorded mule train travel as early as 1853.[31] And Hambly noted the operation of a mule train from Nelson Point to Elizabethtown in 1854. Terwilliger operated a mule train to Nelson Point from Rabbit Creek in 1855. In the same year, J.S. Davis ran a train on the same route,[32] while McClure operated a saddle train from Nelson Point to Rabbit Creek via Gibsonville.[33] Finally, Thompson operated a fourth train in that year on which Haun traveled:

Nov 55
Mo 19 I & Jack went down to the
point prepard to go down to M
ville, we had to wait for Thomp
sons train, we took dinner with
Mrs. Fox. Late we mounted the
horses & drove 58 head of horses
& mules up the hill in company
with Thompson & Read & two
mexicans. We got to onion val
ly after sunset, the wind blew
quite hard & it was verry cold.
We traveld on to grass vally 12
miles moore over the snow & ice
& stopd for the kt.[34]

Courtesy Wells Fargo Museum

Walk-way and bridge at Nelson's Point

Edwards noted the existence of a train route to the American Valley:

> Started for American Valley took express train from Nelson Point arrived at American Ranch 1/2 Past 7[35]

In 1856, Walker & Foots operated a saddle train to Nelson Point and a wagon from Quincy to Elizabethtown.[36]

By 1856, the mail was carried by saddle train:

> U.S. Mail and Passenger Train from Rabbit Creek to American Valley; leaves every morning at 6 o'clock, Gibsonville at 1/2 past 8 o'clock, Onion Valley at 10 o'clock, Nelson Point at 1 o'clock p.m., arriving at American Valley 1/2 past 3 o'clock p.m.

> Returning:—Leaves American Valley every morning at 1/2 past 5 o'clock a.m., Nelson Point at 8 o'clock, Onion Valley at 10 o'clock, Gibsonville at 1 o'clock, arriving at Rabbit Creek in time to connect with the California Stage Companies coaches for Marysville and Sacramento City.
> The train is under the Superintendence of careful and experienced riders, whose whole attention will be given to the accomodation of the traveling public.
> S.T. Brewster, Prop.

> Quincy Hotel—A line of stages leaves the house daily. This house is the headquarters for Brewster's Passenger Train which runs daily to and from the following places; Nelson Point, Onion Valley, Rabbit Town, and the lower cities.[37]

Brewster operated his saddle train as late as 1866, at which time construction began on the La Porte-Quincy Wagon Road:

> Daily Express—Brewster's Saddle Train, from La Porte is making daily trips to Quincy.[38]

By October of 1867 Brewster had switched to stages.

WAGONS AND STAGES.

Travel by wagon or stage depended entirely on the development of roads throughout the region. As late as 1858 Nelson Point was:

> ...absolutely without a road, or a street, nor is there use for one, for there is no way to get a team here, and no room to use one if it was here. The toll-bridge, extending across the two streams which here unite and form the Middle Fork of the Feather River, and, along the farther precipice, some distance, like the shelf over which we rode when coming into town, is a mere foot-walk, slightly made, and covered with split lumber, presenting one of the most perfect nerve-shakers that we ever saw. But in this mountain travel, one becomes so accustomed to every form of danger, (and a good horse learns soon,) that these things are passed almost without a thought.[39]

Accessibility became a major concern of many camps throughout the northern lode. Therefore, as wagon road routes were proposed each town became responsible for convincing the public and the legislature that a route via their town was most advantageous to everyone.

The most significant wagon road route contemplated was one that would connect central Plumas County with the Sacramento Valley. Two routes appeared most feasible—via the North Fork or Middle Fork gorges. The North Fork would favor Quincy while the Middle Fork would establish Nelson Point as the center of commerce in the region. Some questions arose concerning the advantages of the Middle Fork route:

> The general advantages arising from each road, to the people of the county, would be the same. We believe the route along the Middle Fork is shorter than that along the North Fork; but whether it is as easy of construction, is a question in our mind, and could only be determined after a careful survey—hence, we are so anxious that a practical survey of these two routes should be made. This should be done at the expense of the county and we here call the special attention of our Board of Supervisors to its importance.
>
> The road along the Middle Fork would come up to Nelson Point, and, from there to Quincy, a distance of ten miles, a wagon road could easily be constructed: also, from Nelson Point to La Porte, in Sierra County, a good wagon road could be built with little cost. Nelson Point would then become a place of commercial importance, as it would then, during a great portion of the year, be the depot for all this section of the county. From Nelson Point, the main road would likely to come by way of Quincy, or, at any rate, by way of the American Valley, and thence along the route already indicated, to, and through Beckworth's Pass.
>
> This route labors under disadvantages, even if it should, after a survey, be found more feasible than the North Fork route,—for this reason: the latter has more inhabitants living along its contemplated course, and has already attracted the attention of more men of capital and energy than the route down the Middle Fork.
>
> We tell this to the people of Nelson Point and the advocates of the Middle Fork route, that, if they wish to secure a favorable consideration for their route, they must do something that will bring its superior merits, if it has any, before the public mind, or they will have 'sinned away their day of grace," and the North Fork route will bear off the palm.[40]

All of these considerations developed into the Oroville & Beckwourth-Pass Wagon Road project. On June 18, 1866, the certificate of incorporation was filed. Construction of the road from Oroville to Beckwourth-Pass was included in the projections for both routes. However, the project failed, "no work having even been commenced."[41] Between the failure of the road project and declining placer deposits, Nelson Point's fate had been sealed. The development of wagon roads established Quincy as the center of commerce in Plumas County. Nelson Creek was forced to settle for a subordinate role. This lesser role would still rely on the development of another wagon road through Nelson Point.

One proposal was to build a road from American Valley to Sawpit Flat via Nelson Point:

> Any man, claiming to be a philanthropist, would second the building of a wagon road from American Valley. to Sawpit Flat, via Nelson Point. The grunts escaping from mules under heavy burdens are heart-rendering, without the continuous stream of blessings poured on their devoted heads by those attending them. If ox-teams traveled the road, the later pious expressions would become intelligible by being forcibly

Courtesy Jann Garvis

Oroville-Quincy stage, Nicholas Pauly driving

translated from jargon Spanish to Missourian. The luminous idea of facilitating travel and trade with a neighboring county is hard to get through a Board as long as Road Funds are low; but if is a bad time to talk about it, so we will wait till Spring before renewing the subject.[42]

The proposal, however, which would significantly extend the life of Nelson Point was that to build a road from La Porte to Quincy. The state legislature passed the Act of March 31, 1866, authorizing construction of the La Porte-Quincy Wagon Road. E.H. Pierce was hired by the county to construct the road. But after disbursing a considerable sum, the county became discouraged. La Porte bankers Conly & Co. expended $30,000 which exceeded by $10,000 the county bonds issued to finance the road.[43] The project was part of a scheme to annex La Porte to Plumas County. It was also intended to provide American Valley farmers new markets for their products. Unintentionally, the new road would reestablish Nelson Creek as a significant location of commerce.

By October of 1867 the road was in use and the citizens of Nelson Creek were faced with an unavoidable decision:

Oroville–Quincy stage, Jim Whiting on the front seat

> Teams have commenced to travel on the new road. We presume the town authorities will after due consideration of the subject, conclude to remove the town to some point nearer the road.[44]

The La Porte-Quincy Road had been routed several hundred feet above the Nelson Creek townsite and, consequently, a death sentence had been issued for the town at its original location at the confluence. A move was deemed necessary to revitalize Nelson Creek commerce. Nelsonites had no choice other than to relocate the town on the road. In consequence, it was rebuilt at a high point, approximately a quarter-mile from where the road crossed the Middle Fork and nearly a half-mile from its original location. As a result, new opportunities opened for stage and express lines.

Samuel T. Brewster dropped his passenger train business (via the old trail) in favor of a stage line on the new road by October of 1867:

> Stages—A line of stages has been placed on the road between Quincy and La Porte, and will hereafter make regular trips every other day. Dr. S.T. Brewster is the proprieter—mail carrier and stage driver.[45]
> New Stage Line—A line of stages will be placed upon the road between Quincy and La Porte, by Dr. Brewster, making three trips a week, every other day.[46]
> La Porte Stage—Doc. Brewster's stages are now running regularly between Quincy and La Porte.[47]

By December of 1868, Brewster merged with Wilson, adding a route from La Porte to Oroville.[48] Later that month competition with Brewster's enterprise developed when the La Porte-Quincy Express began operations:

> The La Porte-Quincy Express, William H. Knowles, weekly between La Porte, Saw Pit, Onion Valley, Nelson Point, Quincy, during the winter. First trip December 28th, then every Monday.[49]

Knowles specialized in winter travel:

> Knowles' Snow Shoe Express between La Porte, Saw Pit, Nelson Point, and Quincy—Leaves La Porte every Monday morning.
> W.H. Knowles, Proprietor

Brewster could not compete with Knowles' winter operation. Three years later the *Plumas National* wrote:

> Final Insertion—Snow Shoe Mail. We are informed that Mr. William Knowles has been engeged by Dr. Brewster to carry mail weekly between Quincy and La Porte during the winter season.[50]

WINTER TRAVEL.

Nelson Point was regularly isolated by heavy winter snow-fall, consequently limiting mobility. The miners of the region utilized two types of snow shoes as means of overcoming the travel restrictions imposed on them by mother nature. Any mention of snow shoes usually referred to some form of the traditional snow shoe that we know today. However, "norwegian shoe" or "longboards" referred to skis. Norwegian snowshoeing was mostly limited to the higher elevations surrounding Onion Valley, while traditional snow shoe travel was the most common mode of winter transportation at Nelson Creek prior to 1860. Details of Norwegian snowshoeing are offered in the Sports and Postal Service chapters.

Haun made clear the difficulty of winter travel encountered during a trip to Nelson Point:

Traveling by snowshoe, La Porte

Dec 55
We 12 …At 10 oclock I & dave sta
rted for Gibsonville & left wife
& Lizzy at the rate of 14$ a piece
per week. We arrived on foot at
Gibsonville at 1 P.M. the sun shone
warm & pleasant all day.
Th 13 Clowdy this morning. We had
made a pair of snow shoes a piece
& paying our bill for three meals
& lodging each, 6$ at the mount
ain cottage we started for Nel
son point. We had gotten about
one mile & had to return. I forgot
my revolver, besides we could
not keep our snow shoes on.
We again started after 10 A.M.
& with mutch difficulty & tying
we got our shoes to stay on. By

this time the snow begin to fall
& wind to blow. We got to Onion
Valley after noon, waited 1/2 hours
for dinner 2$. Still snowing a
little we tide on our shoes & again
started, the wind blowing a perfect
gale. The snow drifted & light.
We arrivd at the point after sun
set quite fatigued & the skin
rubd off our feet in severall
places, but went on home.[51]

Between the 16th and 20th of that month John Haun—James Haun's son—recorded the preparations for their return to Rabbit Creek:

> ...staid about the cabbin all day & made snoe shoes out of hoops & flour sacks...coming down we wore our snow shoes to try them...Pa Dave & I started for Rabbit creek arrived at onion valley nearly noon got something to eat & fixed our snow shoes & started on...[52]

James Haun continued his account of the return to Rabbit Creek to get his wife and Lizzie:

Dec 55
Su 23 ...we dug up a sledge out of the snow
I went & got a young fir tree &
made half soles for the runners
tide on our luggage & put my
wife on, pd the land lord 10$
off we started 4 of us to pull &
Liz to get along the best she could
as soon as we got in the woods
the snow crest brok through,
so Dave went on to Gib ville af
ter our snow shoes & met us one
& 1/2 miles from Gib ville arrivd at
noon, took dinner 6$ & soon af
ter started for the onion vally, but
got to the foot of the hill & returnd
again to the mountain cottage.
Mo 24 Clear & verry cold. Last kt &
all of to day after 10 a.m. paid
our bill 12$ we set out again wi
th some 3 men to help us & got up

the hill half way, but the wind
blew from the N.E. so cold we was
afraid to risk the top, for fear of
freezing wife & Liz, we turnd about
again & went back to the moun
tain cottage put up for the day
I & John went out & got some tim
ber to make a knew sledge, we
succeeded in getting it partially made.
Tu 25 Christmas morning & cold yet
but moderated some & clowding
up we set two & finished our sledge
& harnessed up. Pd. our bill 16$ off
we started, we had gone 1/2 mile
when 13 men over took us to help
us over the mountain, we halld
them mostly up to it top, some 4
returnd the rest went on to onion
vally, we put up for the kt, the
men took a hour , eat dinner &
returnd to Gibson ville, & snowing.
We 26 Snowd all kt & at this place
the wind blowing a heavy gale with
the snow falling fast, after late
breckfast we hitchd up paid our
bill & those that helpd us the day be
fore 18$ & askd aid to draw our
sledge load through the snow
a part of the way 7 men turnd
out, we had gotten over the har
dest places 3 turnd back & the oth
er 4 went on to Nelson point
arrivd all in good order & in time
we put up at the Thompson
house, it's snowing still...[53]

Edwards also recorded winter travel in 1855:

> Express Messengers arrived over the mountains 2 days from Gibsonville on snow shoes[54]

The difficulty of winter travel induced F.B. Whiting in 1858 to buy three large dogs, Newfoundland and St. Bernard, and he constructed a sled. His subsequent express business became legendary. When snow blocked the stage road, Whiting and his dogs covered a twenty-two mile route between Buckeye and Meadow Valley, effectively connecting Plumas County with the outside world. Whiting discontinued his winter express in 1865 when the horse snow shoe was introduced, enabling stages to travel in the snow.[55]

Fenton "F.B." Whiting's Dog Sled Express

And finally, the California State Agriculture Visiting Committee chronicled winter travel in 1858:

> NELSON POINT.
>
> Crossing another Divide, similar to the one between Gibsonville and Onion Valley, but still higher, and loaded with greater quantities of snow, six miles brought us to Nelson's Point, on the Middle Fork of the Feather River. This last crossing, we think, has fairly initiated us into mountain trail traveling. During the winter, when the snows are deepest, men who are perfectly acquainted with these mountain routes, go out on snoe-shoes take with them each a back-load of shakes, i.e., flat pieces of pine, split out, four feet long, six inches wide, and one-half to three-fourths of an inch thick, a hammer, nails, etc., and passing from tree to tree along the route nail one of these shakes across the body of the tree at the top of the snow. These serve as guides to the stranger, and to them we were frequently indebted for guidance, when every other resource failed us; and we not unfrequently passed them thirty feet above our heads, while we ourselves were riding upon snow to a great depth. The depth to which the winter snows fall in these mountains would astonish any one not acquainted. Our trail over the divide was frequently entirely obliterated, and we could go by no other direction than the foregoing marks on the trees. It also became very steep after descending below the region of snow—simply a mule-path, worn into the crumbling rocks, and going a zig-zag course, turning an angle of one hundred and thirty-five degrees every few yards, and then not unfrequently clambering down cliffs of rocks, which at first seem positively impassable. Having reached a point within one hundred yards of the hotel, the rock was so hard and steep that they had been obliged to build up from below and lay plank in order to get sufficient footing for an animal to stand...[After a visit at F.S. Fox's hotel the committee crossed the foot bridge at the confluence]
>
> From this to American Valley, thirteen miles, we found less snow, but quite a sufficiency of steep climbing until the last five or six miles, when our trail laid along one or the other side of a very sharp ravine, frequently crossing the stream, but maintaining only the moderate descent of the water-course, and a very good trail...[56]

BRIDGES.

Edwards gave the earliest account of a toll bridge at Nelson Point in 1853. The toll was 25 cents.[57] This structure and the bridge described in the previous section on Wagons and Stages were only two of many bridges to have been built to span both Nelson Creek and the Middle Fork at several locations. In May of 1867 H. White of Nelson Point was granted a license by the Plumas County Supervisors to operate a toll bridge across the Middle Fork for six months. By October the Nelson Point bridge was completed and was said to be, "a very superior structure."[58] However, one after another, the bridges succumbed to floods. Area residents became accustomed to

Nelson Point Bridge, circa 1880

such incidents. In January of 1875, for example, because the mail had not arrived from La Porte, Quincy residents surmised that the Middle Fork bridge near Nelson Point had washed out.[59] During the flood of '62 the Nelson Creek Bridge, "a new one just completed," was destroyed. That flood also washed away H. White's timber which was intended for the new Feather River Bridge nearly ready to be raised.[60]

The advent of wagon roads made it necessary to construct reliable bridges. Between 1880 and 1882, $21,661. were spent by Plumas County for bridge construction. Nelson Point was one of four bridges built from that fund. It was an arch-truss bridge constructed by the San Francisco Bridge Company at a cost of $5,787. The bridge spanned the Middle Fork on the La Porte Road.

Later Andrew Jackson Quigley of La Porte built another bridge.[61] And yet another bridge on the La Porte-Quincy Road spanning Nelson Creek was completed in 1890. Near the turn of the century the last two bridges—these made of steel—were constructed on the La Porte Road. One spanned Nelson Creek and the other the Middle Fork.[62]

Transportation was the most critical element of the infrastructure of the gold rush community. Particularly in the northern lode where food and supplies had to be carried from the valleys, an efficient network of transportation was essential to the production of the mines and the development of the region.

RIGHT:
Mrs. V.W. Dempsey and Ruth Haddick standing on the Nelson Creek Bridge in 1918

BELOW:
Nelson Point Bridge, Middle Fork of the Feather River

Courtesy Jann Garvis

NOTES.

1 Fariss and Smith, *History* (San Francisco: 1882), pp. 230-31.
2 Ibid., p. 230.
3 Haun, *Diary* (Plumas County Museum Archives: unpublished), Dec. 6-7, 1854.
4 Ibid., Mar. 22, 1855.
5 Edwards, *Diary* (Bancroft Library: unpublished), vol. 9, Apr. 3, 1855.
6 Haun *Diary*, Apr. 9-10, 1854.
7 Edwards *Diary*, vol. 10, May 31, 1855.
8 Ibid., vol. 14, May 25, 1857.
9 Kimball Webster, *The Goldseekers of '49* (Manchester, N.H.: 1917), pp. 162-63.
10 Haun *Diary*, Sept. 18, 1853.
11 Wooden troughs called flumes were constructed along the streams throughout the region for the purpose of supplying water to the mines.
12 Fariss and Smith *History*, p. 231.
13 California State Agricultural Society, "Report of the California State Agricultural Society," *Transactions: California State Fair and Exposition* (Sacramento: 1858-60), pp. 204-5.
14 Frank Lecouvreur, *From East Prussia to the Golden Gate* (New York: 1906), pp. 225-6.
15 Hambly, *Autobiography* (Bancroft Library: unpublished), Apr. 17, 1854.
16 Edwards *Diary*, vol. 9, Mar. 21, 1855.
17 Ibid., vol. 8, Nov. 12, 1854.
18 *Plumas Standard* 3, no. 34: 2.
19 Fariss and Smith *History*, p. 232. See my chapter entitled Postal Service for a complete explanation of U.S. Postal Service operations in the vicinity.
20 *Plumas National* 2, no. 24: 4.
21 Fariss and Smith *History*, p. 231; E.A. Wiltsee and W.R. Parker, *The Franks of the Everts Expresses*, quoted in Helen W. Gould, *La Porte Scrapbook* (La Porte: 1972), p. 52.
22 Haun *Diary*, Dec. 10, 1854, Feb. 24, 1855.
23 Edwards *Diary*, vol. 9, Feb. 24-5, 1855.
24 Fariss and Smith *History*, p. 216.
25 Fariss and Smith *History*, p. 231; Gould *La Porte Scrapbook*, pp. 54-6.
26 Edwards *Diary*; Plumas County, *Plumas County Merchant Record 1854* (Plumas County Museum Archives: unpublished).
27 Edwards *Diary*, vol. 11, Sept. 24, 1855.
28 Ibid.
29 Haun *Diary*, Feb. 28-9, 1856, May 16-7, 1857.
30 Rabbit Creek was later renamed La Porte.
31 Haun *Diary*, Nov. 14, 1853.
32 Edwards *Diary*, vol. 9, May 3, 1855.
33 Ibid.
34 Haun *Diary*, Nov. 19, 1855.
35 Edwards *Diary*, vol. 10, Aug. 28, 1855.
36 Haun *Diary*, Feb. 22, 1856.

[37] *Plumas Argus*, Apr. 29, 1858.
[38] *Quincy Union*, June 9, 1866.
[39] California State Agricultural Society *Transactions*, pp. 203-04.
[40] Hubert H. Bancroft, *Bancroft Scraps* vol. 3 (Bancroft Library: unpublished), p. 885.
[41] Fariss and Smith *History*, p. 235.
[42] *Quincy Union* 4, no. 6: 4
[43] Fariss and Smith *History*, pp. 235-6.
[44] *Quincy Union* 5, no. 52 (Oct. 26, 1867): 3.
[45] Ibid., vol. 5, no. 52 (Oct. 26, 1867).
[46] *Plumas National*, Nov. 2, 1867.
[47] Ibid., Aug. 29, 1868.
[48] *La Porte Union*, Dec. 19, 1868.
[49] Ibid., Dec. 26, 1868.
[50] *Plumas National*, Dec. 2, 1871.
[51] Haun *Diary*, Dec. 12-3, 1855.
[52] John Haun, *John Haun Diary* (Plumas County Musem Archives: unpublished), Dec. 16-20, 1855.
[53] Haun *Diary*, Dec. 23-6, 1855.
[54] Edwards *Diary*, vol. 8, Jan. 8, 1855.
[55] Fariss and Smith *History*, pp. 232-3.
[56] California State Agricultural Society *Transactions*, pp. 203-4.
[57] Edwards *Diary*, vol. 4, Sept. 25, 1853.
[58] *Feather River Bulletin*, Aug. 11, 1966: C-18, 19.
[59] *Plumas National* 9, no. 18 (Jan. 23, 1875): 3.
[60] *Plumas Standard*, Feb. 1, 1862.
[61] *Sacramento Bee*, July 23, 1947.
[62] These sites are those of the current concrete bridges.

CHAPTER V.

POLITICS.

COUNTY APPORTIONMENT.

Shortly after its discovery, Nelson Point was placed in the Quartz Township, one of two created in Butte County on August 6, 1851. On March 18, 1854, Plumas County was formed from a part of Butte County. Haun was there to record the election that approved the measure:

April 1854
Sat 8 ...Our election came off today for a knew set of county officers & a knew county taken from Bute & calld Plumas. I voted in the after noon & went work.[1]

The Act of March 20, 1855, divided Plumas County into three districts. District No. 1 included Quartz and Washington Townships. As a part of the Washington Township, Nelson Point was placed into District No. 1. John C. Lewis, a Nelsonite, was elected the first supervisor of District No. 1.[2]

A complete reapportionment of Plumas County took place on August 9, 1866, and the county was divided into eight townships. The Goodwin Township was formed in part by the inclusion of the Washington Township. In 1880 the Plumas County Board apportioned all townships into three supervisor districts. District No. 1 included the townships of Quartz, Beckwourth, and Goodwin. George S. McLear became the district representative.[3]

The organization of Plumas County offered Nelsonites local representation and an opportunity to become politically active in the county.

POLITICAL ACTIVITY.

Nelsonites demonstrated interest in local politics from the very beginning. Both Edwards and a man named Spencer acted as poll clerks for the first county election of officers. In his diary Edwards recorded the election results:

	Democratic	Whig
County Judge	Ward............34	Bradley.........27
Prosecuting Attorney	Cox..............34	Barnett 28
Clerk	Lewis 40	Harberson 23
Treasurer	Cate 35	Judkins 26
Sheriff	Kingsbury.....31	Sharp............32
Coroner	O'Niel 35	Vaughan 27
Assessor	Porter...........34	Buckbee 28
Surveyor	Taylor 21	Whitlock......38 [4]

Examining the list reveals that a significant number of the candidates were Nelsonites; subsequent elections attracted similar involvement by Nelsonites. In 1856 Edwards ran for Justice of the Peace and recorded the results by precinct:

	Edwards	Stevens
Rich Bar	50 votes	3
Independence Bar	30	3
Nelson Point	35	31 [5]

> ...Sherwin returned from Quincy & informed me I was elected Justice of the Peace for Washington Township[6]

In the same year, Edwards noted that J.L.C. Sherwin and J. L. Lewis, both Nelsonites, were elected president and secretary respectively of the Democratic County Convention at Quincy.[7]

Edwards remained active in the Democratic party. In June of 1857 he:

> Wrote and posted 3 notices calling a meeting of the Democracy of Nelsons Point precinct on Saturday 13th inst to elect delegates to Democratic County Convention to be held June 20th at Quincy for the purpose of nominating County Officers & Delegates to Democratic State Convention[8]

Having been elected as secretary at the meeting, Edwards recorded the event:

> Evening Agreeable to notice the Democrats of Nelsons Point precinct assembled at Thompson House to elect Delegates to Democratic County Convention to be held at Quincy 20th inst for the purpose of nominating County Officers & to elect Delegates to attend the State Democratic Convention
>
> I called the meeting to order & nominated J.L.C. Sherwin for president who was unanimously elected & who after taking the chair nominated me for Secretary & I was elected.
>
> We then elected John A. Ritchie & James H. McNabb as said Delegates & Lewis Cross as substitutes in case of the non attendance of the regular Delegates elect & to whom I delivered credentials[9]

It is clear from election results in the early years that a small group of men was primarily responsible for political activity at Nelson Creek and that, even though relatively limited, the following they enjoyed was sufficient to elect them. The election vote tabulations also suggest that the vast majority of men at Nelson Creek did not vote. There are two likely explanations for this. The first is that the miners were too busy mining to vote. After making their fortune they planned to return east and, consequently, felt no desire to become involved in domestic matters at a place they did not consider home. Secondly, some evidence suggests that low voter turnouts were due in part to the fact that those who could not prove citizenship were not allowed to vote. Edwards witnessed this at Independence Bar:

> Being election day we quit work a little earlier than usual & went to Independance Bar to deposit our votes several persons informed me that the Judges refused to accept the vote of an old Frenchman (commonly known as Yankee Jack) because his naturalization papers were not issued by the District Court but by the County Court of Plumas therefore contended they were illegal...he wished to deposit and insisted the vote be polled, on which the old man came forward the form of oath was read & interpreted to him he was sworn on the Mountain Messenger the organ of the Know Nothing party of which the Judges were members & his vote polled
>
> M^c^Guyre was also challenged and swore in his vote on the same copy of (holywrite) "in a horn")[10]

Haun was intrigued by local politics as well. Unlike Edwards, though, he played a more passive role. John noted his father's political affiliation in August of 1855, "...all of us went down to the point after supper Pa joined the Know Nothing party..."[11] At every opportunity Haun kept informed regarding local political matters. For example:

Sept 55
Tu 4 ...After supper Roister & I
went down to the point we
had four political speeches
& good order something uncommon
the speaking over we put for home.
We 5 ...P.M. we three went down to
the point to the Election...[12]
June 58
Fr 18 ...The Boys went down to the point
I hurid on my best duds & made
for the vally I arrived at sunset
quite a number was in attendance

on the county convention
Sat 19 ...quincy was full of men to day,
the convention nominated all
Bucannon Candidates for office
the Douglass men were all Beaten.
by a coalition of Know nothings
old whigs & Bl Republicans
another show tokt. &
Great dissatisfaction among
the duglass democratic ranks[13]

One controversy was responsible for eliminating most indifference to politics throughout the mother lode—secession. The Civil War became a heated political topic in the region. Quincy developed into a Union Party stronghold and it appears that Nelson Point favored the Union as well. In 1861 a "Union MEETING AT NELSON POINT" was reported. The meeting was held on May 12th and a "Union Club" was formed. The club's purpose was to uphold the Union and to insist that those who claimed to be out of the Union were in fact not out and that they obeyed the laws of the United States.[14] Nelson Point was entitled to two delegates at the 1865 Union Party Convention in Quincy.[15]

Newspapers of the time were often politically oriented. The *Quincy Union*, for example, became a "party paper" for the Union Party and the *Plumas Standard* offered a Democratic Party opinion.

Perhaps the best measure of interest in politics during the later years at Nelson Point was the number of registered voters. *The Great Register of Plumas County* listed the following voter registrations for Nelson Point:

August, 1875 19
October, 1884 38
1892 13

Results from the *Precinct Registers of Plumas County* after the turn of the century continued to be recorded under the "Nelson Point Precinct." Votes recorded in the 1900, 1912, 1924, and 1928 numbered as follows:

1900 31
1912 7
1924 14
1928 2

For a more detailed examination, the precinct registers are included in Appendix XI.

In 1885 Hambly reminisced about the political activity in the region during the early years:

> ...from this date a few little oddities and "pickerys" can be culled from a compilation of events written up & recorded in the so called History of Plumas, Sierra & Modoc Cos. but discrepances are rife, in the whole thing, there is too much ring of money & shoddyism in it, as regards, primitive scintillations & wisdom of the Law Makers of Plumas I have nothing to find fault with; chicanery with a chance to handle the "dear peoples" money is 99/100 per cent of the desires of politicians my opinion of the greed of money blamed in Esq Bonners Court could be winked at, in subsequent Courts & open handed stealing allowed under the cover of Law & Order. but my statements here in this matter is neither Pro or Con in the future working of Plumas as a County organization in the enlightened age we are doing well, one reason I expect is, our political proclivities are pretty well balanced, & the parties in power, knows full well there retaining office, depends on their honesty & ability to serve the dear people, but the best men are not always elected to office, wire pulling, & button holing, has become a science in the hands of political tricksters & sycophants of the money power & corporations, we have been very unfortunate in our Rep.s to Sac City up to this date, the Rep side of the house especially. Tools of the recreant but more wiliy political intrigues, the last spawn of such is the "debris question" is not a pity and a loss to the people that our incumbant as a P.M.: was not sent, his illustrious record ought to keep him in the field as a bright luminary of the Rep. comments not necessary.[16]

Political activity at Nelson Creek was a mixed bag of enthusiasts, the curious, and the indifferent. Ironically, the evidence suggests that a significant number of the county's most prominent politicians originated from Nelson Creek and that they played an essential role in the development of Plumas County.

NOTES.

1 Haun, *Diary* (Plumas County Museum Archives: unpublished), Apr. 8, 1854.
2 Fariss and Smith, *History* (San Francisco: 1882), p. 161.
3 Ibid., p. 166.
4 Edwards, *Diary* (Bancroft Library: unpublished), 1854.
5 Ibid., vol. 14, Nov. 4, 1856.
6 Ibid., vol. 14, Nov. 15, 1856.
7 Ibid., vol. 14, Sept. 1, 1856.
8 Ibid., vol. 14, June 7, 1857.
9 Ibid., vol. 15, June 13, 1857.
10 Ibid., vol. 10, Sept. 5, 1855.
11 John Haun, *John Haun Diary* (Plumas County Museum Archives: unpublished), Aug. 11, 1855.

[12] Haun *Diary*, Sept. 4-5, 1855.
[13] Ibid., June 18-19, 1858.
[14] *Plumas Standard* 2, no. 34: 3.
[15] *Quincy Union* 3, no. 34: 2.
[16] Hambly, *Autobiography* (Bancroft Library: unpublished), 1885.

CHAPTER VI.

NELSONITES.

> ...nothing *that glitters is gold. So I learned then, once and for all, that gold in its native state is but dull, unornamental stuff, and that only low-born metals excite the admiration of the ignorant with an ostentatious glitter. However, like the rest of the world, I still go on underrating men of gold and glorifying men of mica. Commonplace human nature cannot rise above that.*
>
> —Mark Twain

The population of California skyrocketed during the early gold rush years; by 1860 it had swelled to 380,000. The population of Plumas County at that time was 4554, of which 279 resided in the Washington Township. The Washington Township, under which Nelson Point was recorded in the census, categorized the races as follows:

185 white
80 Chinese
4 Mexican
0 black[1]

It is impossible to determine precisely the population of Nelson Point. Numbers constantly fluctuated as miners came and went on a daily basis. The best estimate is that at any one time the total numbered five hundred or more. The population exploded in 1850, peaked in the mid-50's, and had become largely Chinese by the early 1860s. By 1890 the Chinese had left and the population had dwindled to seventy-five.[2]

The Plumas County registers of Nelson Point voters indicate the extent and timing of the population decline in its latter days:

1875 19
1884 38
1892 13
1900 31
1912 2
1924 1
1928 0

The gold rush attracted men of all social classes and educational backgrounds. The northern mining camp contained a representative sample of men from all over the world. Perhaps their only common characteristic was their willingness to gamble on their own futures. At nearby Rich Bar, Dame Shirley characterized the men:

> Perhaps you would like to know what class of men is most numerous in the mines. As well as I can judge, there are upon this river as many foreigners as Americans. The former, with a few exceptions, are extremely ignorant and degraded; though we have the pleasure of being acquainted with three or four Spaniards of the highest education and accomplishments. Of the Americans, the majority are of the better class of mechanics. Next to these, in number, are the sailors and the farmers. There are a few merchants and steamboat-clerks, three or four physicians, and one lawyer. We have no ministers, though fourteen miles from here there is a "Rancho," kept by a man of distinguished appearance, an accomplished monte-dealer and horse-jockey, who is said to have been—in the States—a preacher of the Gospel. I know not if this be true; but at any rate, such things are not uncommon in California.[3]

Nelson Point boasted many of the most successful and distinguished men in the county. Brief descriptions of the most noteworthy follow.

NELSON—was credited with the discovery of Nelson Creek. He was a "tall, sandy haired, thin faced, good natured hombre,"[4] who mined and operated a store at the confluence of the creek and river.[5]

JOHN WESLEY THOMPSON—was born July 1, 1829, in Bethel, Ohio. He came to California via Panama in March, 1850. In November, 1850, he arrived at Nelson Point where he built a log cabin and opened his trading post. Thompson made a fortune mining, freighting, and operating the Thompson House and the Illinois Ranch, which he purchased in 1853. In 1858 he sold the Thompson House and returned to Ohio. In the early 1860's he returned to the region to operate the Illinois Ranch where he lived with his wife until his death in 1905.[6]

RIPLEY C. KELLEY—came to Nelson Creek in September of 1850 where he joined his friends in the Wisconsin Company. He returned to Wisconsin in the winter with substantial profits. Later, he came back to Plumas County and was elected assemblyman in 1855. He resumed mining shortly thereafter mining at Poorman's Creek as late as 1882.[7]

JAMES L. C. SHERWIN—without question became Nelson Creek's most prominent citizen. He came to Nelson Point in 1850 where he formed a partnership with J. W. Thompson. Later he operated the Sherwin House at Independence Bar. Sherwin diversified his operations at Nelson Creek, becoming best known for his flume (water company) and "Sherwin's Mill." Sherwin acted as Nelson Creek's only known pathmaster. And he was a conspicuous member of the Vigilance Committee. In 1857, Sherwin was unanimously elected as president of the Democratic County Convention Delegates. Correspondence in the *Plumas Standard* fondly remembered Sherwin's contributions to Nelson Point:

> ...I believe my authority will abundantly justify me in saying that J.L.C. was a great man. He was an active, bustling sort of man, and positively the greatest talker that ever lived. Mr. Peter Newsby, in speaking of the loquacious tendencies of this great man once said, in my presence, that J.L.C. always put him in mind of a bumble-bee. But let us be a little more serious for once—Sherwin was enterprising, ambitious, and full of romantic schemes. He quit merchandising in '53, to engage in a mining enterprise. By his indomitable energy in the spring of '54 two large flumes sprung up like magic. They conveyed water to the miners on the bars and flats of the creek and river for a distance of about three miles. The flumes cost about $25,000., were only moderately good property, but a perfect God-send to the people of Nelson. He represented our county in the Assembly, during the session of 1857, and was honest and faithful. That you will admit is saying a great deal for a California Assemblyman. Sherwin left his mark on Nelson Point, as indeed he will do wherever he may wander. Many laughed at his road scheme, and more, perhaps, at his political course; but we had better all admit at once that the road he worked so hard to get through is sorely needed at this time...Had we had some more men in the county of Sherwin's public spirit, it would have been a great deal better for us.[8]

Sherwin left the county several years after his legislative experience In 1867, he advertised "a large amount of valuable property, at Nelson Point for sale."[9] Sherwin never returned.

Frank S. Fox—was in the early 50's the cook and operator for Thompson at the Thompson House. He bought out Thompson in 1858. Mr. and Mrs. Fox operated the place until the early 60's when they experienced financial difficulties.

Major James H. Whitlock—was born in 1829 in Union County, Illinois. He arrived at Nelson Creek in March of 1851. There he mined and supplied lumber for flumes, but met with little success. In 1852 he and his partners, the Hargraves, operated a claim at Fiddler's Flat. He went below during the winter of 1853 and returned to his diggings in April. Whitlock defended Mr. Gilson in court during the Gilson-Wilson trial. In the spring of the following year he was elected Plumas County Surveyor and was re-elected for four terms.[10] Venturing to Arizona, he fought in the Apache Indian Wars where he earned in 1864 the rank of major. Major Whitlock returned to Quincy in 1877 where he remained active in politics and business.[11]

James Humphrey Haun—departed from Georgetown, Kentucky, at the age of 41, and traveled to California via the isthmus in the spring of 1853. He arrived at Nelson Creek on June 21, 1853, where he made a fortune in various mining enterprises. Shortly after his family's arrival in 1855, he bought the American Ranch in Quincy. There he operated one of the first hay, grain and vegetable farms in the valley, while at the same time overseeing his mining interests back at Nelson Point.

Courtesy Fariss & Smith

Major James H. Whitlock

He and his wife opened at the ranch one of the first hotel-taverns in the valley as well. He died in 1890. Haun kept a diary while at Nelson Point in which he recorded a vivid account of gold rush life.[12]

JOHN HAUN—the son of James. He followed his father to Nelson Creek and mined at the family's diggings there. Like his father, John kept a diary. Unfortunately, the volumes prior to January of 1854 were destroyed when his cabin burned. Though not nearly as descriptive as his father's diary, John's writings offer valuable supplementary information.

DAVID WILLIAM HAMBLY—was born in England on October 5, 1820. He arrived in the United States in 1844, originally settling in the Wisconsin Territory. Hambly departed for California on March 12, 1850, traveling overland. He arrived first at Nelson Creek in 1852 in search of his brother, William, whom he later found in Grass Valley. Upon his return to Nelson Creek in the fall of 1852, he and his brother bought an interest in the Buckeye Claim. Having experienced moderate success at mining, Hambly relocated to Snake Lake. In 1885 he wrote his autobiography which he compiled from the diary that he kept during the years he lived in the region. It offers a more humanistic view of life at the mines than any other account discovered to date. In his later years at Snake Lake, Hambly's Humanism turned to Spiritualism. He died in 1890.

WILLIAM A.Z. EDWARDS—Arrived in California via Panama in the early 1850's. By 1853 he resided at Nelson Creek where he mined and became prominent in local politics. Edwards became an active member of the Vigilance Committee as well. In 1855, he abandoned mining and moved to Poplar Bar on the Middle Fork where he supplied beef to the miners at Nelson Creek. By early 1857 he returned to Nelson Creek to resume mining, and was elected the same year as the Justice of the Peace for the Washington Township. He became a prominent socialite at Nelson Creek, rarely missing any public gathering. For this reason and because of his extensive political involvement, his diary offers rare details of life at Nelson Creek during the early years. By July of 1857 Edwards had moved to the bay area where he raised mustard seed.

James H. Larison—was a native of Ohio and came to California via Panama in 1850. He returned to Ohio for his family in 1852. James and his family arrived at Nelson Creek in 1853, where he mined until 1866. In 1854 he boarded Edwards. Two years later he mined with Terwilleger & son and was a juryman in the Haun-Lloyd dispute. In 1866, he bought the Boynton Ranch in the American Valley and renamed it the Kentucky Ranch. There he produced vegetables. He died of "the grippe" in 1908 at the age of 80.[13]

James Haun

William Clinch—was a native of England who departed for California in 1852. He traveled around the horn, but shipwrecked off the coast of Chile. William worked in the silver mines there until 1854 when he sailed for San Francisco. From San Francisco he made his way directly to Nelson Creek where he mined. He later moved to Sawpit Flat.[14]

James Cooksey—was a native of Linnville, Illinois who, at the age of seventeen, traveled overland and arrived in Placerville in 1852. In March of 1854, he arrived at Nelson Creek where he mined until 1867 when he left for Sierra County.[15]

Nicholas Oscar Pauly—was a native of Germany who, after brief stays in New York and Missouri, traveled overland to California in the 1850's. He became an express agent and traveled from Marysville to the mines. About 1875 he settled at Nelson Point where he and his wife operated the Pauly Hotel. Simultaneously, he operated a freight operation on the La Porte Road. He died in 1903 at the age of seventy-five.[16]

Catherine Pauly—was a young lady in Germany of just eighteen when her uncle, Nicholas Gard, sent passage for both her and her cousin to come to America. Catherine and her cousin traveled six months around the horn to San Francisco. She stayed with the Gards in Browns Valley where she met Nicholas Oscar Pauly. They were married in La Porte in 1871. [The Pauly Marriage License appears in Appendix

Courtesy Jann Garvis

Mrs. Catherine Pauly

Nelsonites at the Pauly Hotel

XIII.] Catherine operated the Pauly Hotel and raised four sons and a daughter, while her husband pursued his other business interests. Following her husband's death in 1903, Catherine maintained the hotel until it burned in 1924. Mrs. Pauly was a fixture at the landmark hotel for nearly fifty years. Up every morning by three or four, she began preparing meals for the stage travelers and miners who patronized the hotel. Roy Post, who frequented the hotel after the turn of the century, fondly remembered her, "Well, she was a great German woman...good cook, loved to talk."[17]

The list of those who frequented Nelson Point reads like a "Who's Who in Plumas County," and, consequently, it would be impractical to write about each of them in detail. However, a list of those known to have appeared in the vicinity is offered in Appendix VII.

Nelsonites were among the hardiest of the rugged class of pioneers in the region, and they developed a reputation for being a rather rowdy bunch during the early years. However, by the 1860's the population had transformed not only by ethnicity, but in character as well. Hambly described this transformation, providing a final evaluation of the character of the people:

> ...there was a very turbulent element of humanity in and around Nelson Point at this time [1853], Ste Bass, Buckley, Vaughn and a few others made old Nelson howl at times.- alas. They are gone a new era has dawned on the sons of men and the change in 30 years are very perceptable to the quiet observer &c about this time, the fall politics was rife the two parties had commenced to buy favor from the "mud sill" element whiskey, the drug so easily, used to inflame the passions of men for good or evil and always attainable by politicians to the sorrow of many, and the history[18] in the organizing of our County could tell some rough doings; if compilers &c recorders would only be honest in their writings of events &c, the discordant elements of by gones, would remain buried, but postion and mammon gives shades, where they do not exist; and honor & virtue too often left in the shade, unchronicled because of, the mudsill element, I will not unearth the dastardly doings of many, that are yet citizens of terra-firma...[19]

NOTES.

[1] Plumas County Census 1860 (Plumas County Museum Archives).

[2] State Resources 11, no. 4 (Oct. and Nov., 1890).

[3] Louise Clappe ("Dame Shirley"), *The Shirley Letters* (Santa Barbara: 1970), p. 124 (hereafter cited as Shirley, *Letters*).

[4] Alonzo Delano, *Alonzo Delano's California Correspondence* (Sacramento: 1952), p 19.

[5] Nelson's first name is unknown at this time. However, it should be noted that a relative of an Ebner B. Nelson recently claimed that Ebner is the Nelson after which Nelson Creek was named. After an extensive investigation I concluded that the evidence does not even place Ebner in Nelson Point. Only proper documentation can substantiate this claim and place credit where credit is due.

[6] Thompson file, Plumas County Museum Archives.

[7] Fariss and Smith, *History* (San Francisco: 1882), p. 197.

[8] *Plumas Standard*, Aug. 2, 1862: 2.

[9] *Quincy Union* 5, no. 37 (Jan. 12, 1867): 3.

[10] Edwards recorded this election in the spring rather than the fall and indicated that Whitlock ran under the Whig Party.

[11] Fariss and Smith *History*, pp. 197-98.

[12] Haun *Diary*; Haun file, Plumas County Museum Archives.

[13] Huskinson, *Huskinson Scrapbook* (Plumas County Museum Archives: unpublished), p. 34.

[14] Ibid., p. 96.

[15] Fariss and Smith *History*, pp. 310-11.

[16] Pauly file, Plumas County Museum Archives.

[17] Ibid.; Leroy Post, interview held at his home, La Porte, California, July 24, 1987.

[18] Hambly referred to Fariss and Smith's *History*.

[19] Hambly *Autobiography*, 1885.

Photo by Scott Lawson

The only remaining gravestone
at the Nelson Point graveyard
at Independence Bar:
"In Memory of Andrew J. Brineger
from near Mineral Pt. Wisc.
who died Nov. 4, 1851.
Aged 27 years."

CHAPTER VII.

FOREIGNERS.

We labor under great disadvantages, in judgement of foreigners.

—Dame Shirley

The California gold rush triggered the first worldwide mass migration in history. Ships from distant ports brought scores of foreigners to San Francisco. They numbered Peruvians, Columbians, Mexicans, Chileans, Chinese, Germans, Frenchmen, Englishmen, Irishmen, and an assortment of other Europeans. The influx of two of these groups is particularly noteworthy—Hispanics and Chinese.

Tracing the contributions that each of these groups made to gold rush history is difficult for several reasons. Our knowledge of the gold rush has to date been exclusively supplied to us by the records, diaries, and journals of whites, who often remained in the region and established new communities in Plumas County. More has been discovered in recent decades by researchers who have located the diaries of miners who returned east following the rush. However, to date scholars have primarily searched the records of white miners' for two obvious reasons. Most Hispanics and Chinese returned home following the gold rush and, consequently, their records are now primarily in Mexico, Chile, and China. Additionally, few researchers have been fluent in these languages. All this adds up to the fact that our knowledge of these peoples is largely based on accounts written by whites who seldom associated with and ultimately lacked understanding of the cultures of these foreigners.

Early in the rush, Hispanics and Chinese worked side by side with the white miners. However, as placer deposits played out an attitude developed among white miners that the gold in America was for Americans. Strangely enough, in this case Europeans qualified as Americans. Consequently, Hispanics and Chinese were ostracized and often lived and worked in communities established adjacent to, but separate from, white establishments.

HISPANICS.

The Mexican-American War, and the subsequent Treaty of Guadalupe Hidalgo in 1848, awarded the Californios to the United States. Mexican rule in California ended, but its influence remained strong. Ironically, just days prior to the signing of the treaty James Marshall made his discovery and soon the rush was on. Together with immigrants from all over the world came Mexicans, adding to the number of Mexicans living in California prior to the rush. As many as 20,000 entered California searching for gold between 1849 and 1851. Chileans came in significant numbers as well.

The Mexicans and Chileans brought with them something that was rare in the early gold rush days—mining experience. Whites, who often possessed little or no mining experience, were eager to learn from the Mexicans. And the Mexicans were usually willing to teach them. But shortly after the rush started attitudes began to change.

As the population boomed and competition for prime diggings increased, a "pecking order" was established in which whites controlled the mines. Friction between the groups developed. White attitudes influenced by the notion of "manifest destiny," emerged. The white proclamation that it was their divine right to spread their civilization across North America generated much of the discrimination experienced by Hispanics and Chinese and in many cases resulted in outright theft of claims from them.

On April 13, 1850, The California state legislature passed the "Foreign Miner's Tax." It required all non-citizens of the United States to pay a fee of $20 per month in order to work the mines. The law became purely discriminatory: it was enforced exclusively on Hispanics and Chinese. On March 14, 1851, the tax was repealed, but the feelings remained unchanged and, consequently, in 1852 the tax was revived but subsequently adjusted downward to $4 per month.

Hispanics accounted for approximately ten percent of California's population between 1854 and 1865. In some locations whites were more tolerant of Hispanics than in others, but generally intolerance prevailed. Hispanics were stereotyped as ignorant, dishonest, and primitive. However, it was more commonly the whites' failure to understand the language and customs of Hispanic culture that fueled their hostility. Dame Shirley related an incident which amply illustrates this point:

> From our Log Cabin, Indian Bar,
> March 15, 1852
>
> ...By the way, speaking of languages, nothing is more amusing, than to observe the different styles, in which the generality of the Americans talk "at" the unfortunate Spaniard. In the first place, many of them really believe, that when they have learned "sabe" and "vamos", (two words which they seldom use in the right place,) "poco tiempo," "si," and "bueno," (the last they will persist in pronouncing "whayno,") they have the whole glorious Castilian at their toungue's end... The majority, however, place a most beautiful and touching faith in broken English, and when they murder it, with the

few words of Castilian quoted above, are firmly convinced, that it is nothing but their "ugly dispositions" which makes the Spaniards pretend not to understand them.[1]

Mistreatment of Hispanics at Nelson Point was not recorded. However, nothing suggests that Nelson Point was any different in this respect from nearby Rich Bar where half the early population was comprised of "foreigners." Dame Shirley noted:

From our Log Cabin, Indian Bar,
July 5, 1852

...About five o'clock we arrived at home just in time to hear some noisy shouts of "Down with the Spaniards;" "The Great American People forever," and other similar cries, evident signs of quite a spirited fight between two parties, which was, in reality, taking place at the moment. Seven or eight of the "elite" of Rich Bar, drunk with whiskey and patriotism, were the principal actors in this unhappy affair, which resulted in serious injury to two or three Spaniards. For some time past, there has been a gradually increasing state of bad feelings exhibited by our countrymen (increased, we fancy, by the ill-treatment which our Consul received the other day at Acapulco,) towards foreigners. In this affair, our countrymen were principally to blame, or, rather should I say, Sir Barley Corn was to blame, for many of the ring leaders are fine young men, who, when sober, are decidely friendly to the Spaniards. It is feared that this will not be the end of the fracas, though the more intelligent foreigners as well as the judicious Americans, are making every effort to promote kindly feelings between the two nations. This will be very difficult, on account of the ignorant prejudices of the low-bred, which class are a large proportion of both parties.

It is very common to hear vulgar Yankees say of the Spaniards, "Oh, they are half-civilized black men!" these unjust expressions naturally irritate the latter, many of whom are highly educated gentlemen of the most refined and cultivated manners. We labor under great disadvantages, in the judgement of foreigners.[2]

Conflicting evidence makes it difficult to determine the numbers of Hispanic miners on Nelson Creek during the early years of the rush. However, it is clear that Hispanics were extensively involved in the region's transportation industry. Many were expert horsemen and their contribution as muleros was invaluable to white entrepreneurs, like Thompson, who operated pack and saddle-trains. In any case, the Hispanic population at Nelson Point in 1853 was apparently sufficient to prompt Edwards to study the Spanish language.[3] Yet through 1854, Edwards failed to mention any Hispanic miners in his diary, nor did either Haun's diary or Hambly's. In 1854 John Haun did remark that, "...I saw the spaniard they had in confinement for killing some Chinamen..."[4] And in 1855 Edwards first mentioned Mexicans on his approach to Nelson Point from Illinois Ranch:

...with the trail made good time passed a couple of Mexicans (camped out beside a roaring fire) with whom I exchanged the usual salutations...[5]

Only with further research can the full extent of Hispanic activities at Nelson Point be known.

CHINESE.

The most misunderstood segment of the gold rush population was the Chinese. They kept to themselves, seldom interacting with the white population. If they kept records, few have survived. The Chinese simply came and went. The elderly returned to China to die, and the younger survivors of the rush relocated in northern California Chinatowns or returned to China. However, their mysterious existence left an indelible mark on the gold rush.

Commonly referred to as "Celestials" because they believed they were descendants of the gods, many Chinese fled the tyrannical rule in their homeland. An estimated 25,000 emigrated to California by the end of 1852.[6] Like most other miners, few brought their families. In 1860, approximately 33,100 Chinese males, but only 1800 females, lived in California.

When they arrived in the mother lode, they congregated outside of white settlements often in sufficient numbers to establish their own "Chinatowns." These Chinatowns, which "respectable" white men did not frequent, operated independently of the larger society's law.

The region's most noteworthy Chinatown was Silver Creek in Meadow Valley. Chinese initially settled at the location after whites had abandoned the diggings. The Chinese profited by reworking the bedrock, which had not been thoroughly worked by the whites. Later, over 200 more Chinese arrived and helped to establish a thriving community. Silver Creek established the only known Joss House (worship center) in Plumas County and, consequently, became a mecca for Chinese throughout the county.[7]

For the most part, the Chinese population of the region kept to itself. Chinese operated their own stores, hotels, and saloons. And like their white counterparts, Chinese miners often displayed an affinity for gambling—Fan-Tan and Chinese Lottery being the games of choice. Whites were rarely found in Chinatowns; those who were, usually comprised the outcasts from white camps.

The Chinese were expert miners. They often mined diggings that had been abandoned by white miners. By thoroughly cleaning the bedrock it is estimated that the Chinese often took out more gold from a claim than had originally been taken by the whites. The Chinese often utilized "wing-dams" to divert the water in order to completely expose the bedrock. The bedrock was meticulously worked with long fingernails, bamboo brushes, and crevicing tools.

Many whites felt threatened by the presence of Chinese miners. Consequently, like the Hispanics the latter were the objects of ridicule and hostility. Three factors best explain why. First, was the mystery surrounding the Chinese customs, language, and appearance. Second, the Chinese became a significant workforce which produced much labor for a fraction of normal wages. And last was manifest destiny. Americans often felt threatened, and reacted by passing a series of laws which discriminated against the Chinese:

- •Foreign Miner's Tax of 1850
- •Revised Foreign Miner's Tax of 1852
- •A law that taxed shipmasters $5. to $10. per passenger
- •Cubic Air Ordinance
- •Queue Ordinance
- •A law which made any corporate employer guilty of a misdemeanor if he hired Chinese (1879)
- •Racial exclusionary rule (1854)
- •Exclusion Law of 1876
- •1882 Exclusion Act
- •Between 1882-1924 the government passed 14 additional pieces of legislation concerning Chinese exclusion laws[8]

California's "Racial Exclusionary Rule" is perhaps the best example of how discriminatory the legislation became. It resulted from the 1854 California Supreme Court decision, "People v. Hall." In that case a white defendant was convicted of murder on the basis of testimony from Chinese witnesses. His conviction was overturned when he asserted that Chinese testimony against a white man was inadmissible by criminal statute. The statute, Sec. 14 of California's 1850 Act Concerning Crimes and Punishments read: "No black or muletto person, or Indian shall be permitted to give evidence on favor of, or against, any white person."[9] California's law was among the most discriminatory of any state. The law was enacted when the courts questioned whether a Chinese witness could be bound by an oath to a God that he did not believe existed.[10]

The Foreign Miner's Tax was collected regularly in 1856 at Nelson Creek on the Chinese at a rate of $4 per month. John Haun documented the practice:

> ...a couple of Chinamen set in today to work for us...the sheriff came to the diggings about noon & got 8$ from the chinamen for taxes...[11]

At one time, Plumas County tax revenues consisted primarily of funds collected under the second Foreign Miner's Tax which was collected exclusively from the Chinese. Of the 911 Chinese adults recorded in the 1870 Plumas County Census, only 24 appeared on the county tax assessment list. They paid assessment taxes totalling $221.08—an average of 24 1/2 cents each. The average white paid $32.14.[12] Consequently, the county felt justified collecting the Foreign Miner's Tax. Twenty percent of the Foreign Miner's Tax was kept by the tax collector (sheriff) for, "twisting the Chinamen's pigtails long enough to collect it."[13]

An 1864 newspaper article reflected public sentiment in Plumas County regarding the tax:

> It Works Well.—The enforcement of the Foreign Mining Tax is operating well in this county. Some of them who "don't like the blaisted country, no how," will have to skedaddle if they expect to get ride of paying the tax. It don't happen to be the Enrolling officer this time. The boot is on the other leg. Don't you see?[14]

Legislation was only one manifestation of white bigotry. More evident were the attitudes displayed routinely by white miners. Much of this was generated by the customs, language, and appearance of the Chinese. Accenting their physical differences was the Chinese attire. Each night after a soap and hot-water bath they routinely donned a complete change of clothes consisting of cotton undershirt, undershorts, and socks. Additionally, dark blue denim overalls with matching jumper were worn. While working, they wore broad rattan plaited hats as well. On dress occasions flat low-crowned felt hats, distinctive silk blouses and matching pantaloons were worn. The most distinctive aspect of their appearance was the queue (pigtail) which was worn on an otherwise clean-shaven head. Queues were particularly well maintained because without them they could not return to China.[15]

Edwards related an experience which occurred while he was walking at dusk along Nelson Creek. It vividly illustrates the mystery surrounding Chinese customs:

> ...as I turned the point of the hill, opposite Henpecked Flatt I discovered a light ahead which I supposed to be persons traveling on the Sherwin Flume but as I advanced I discovered to my small surprise that said light was not on the flume but farther up the mountain side and on coming abreast of it I found the grave of a Chinese who had been accidentally killed on Fiddlers Flatt whilst at work & buried there the grave was nicely fenced around inside of which a board was erected bearing an inscription in chinese characters on either side which was an adamantine candle burning and a tin cup settling near containing tea not yet cold and a spoon in it.[16]

No statistics exist to indicate the extent of the Chinese population at Nelson Creek in any given year, but one estimate puts that populace at perhaps four or five hundred at its peak.[17] A significant number of these concentrated at Independence Bar. Chinese influence was well documented up and down the creek.

"Chinese Slide," a location approximately 1/4 miles upstream from the current Nelson Creek bridge, was named for the Chinese who died when a slide dropped on him.[18] Whether it was the same slide noted by Edwards in the following incident in 1855 is unclear:

> Potts reported that 2 Chinese killed by an Indian with arrows on the trail between Nelson Point & Onion Valley on Saturday Also 1 Chinese killed (by the caving of the bank under which they were working on Fiddlers Flat) & 1 it was supposed would not survive to day & two others seriously injured on yesterday[19]

Edwards recorded the death of another Chinese in 1854:

> ...P.M. Attended meeting of the miners to investigate the case of Page, Tribe, Lane, & White & another in regard to the falling of a tree which killed a chinaman...[20]

There are some references regarding the Chinese mining activities at Nelson Creek. For example, in 1855 John Haun noted:

> we went...to Sterlings diggings & found some chinamen at work in the ravine...[21]

The *Quincy Union* reported the demographic transition at Nelson Creek which was already underway by 1865: as the white population declined, it was being replaced by Chinese:

> Nelson Creek is rapidly passing into the hands of the Celestials. I heard of but four or five companies of white men intending to work the Creek the coming summer.[22]

And in 1867 the *Union* reported:

> The Chinese companies in the river are making money this season, at least dust seems to be plenty with them this season, and they are living on the fat of the land- (hog meat and chickens.)[23]

Both Haun and Edwards make reference to the bigotry aimed at the Chinese. Haun described the destruction of a Chinese house by a Nelson Creek mob:

> Aug 54
> Tu 29 ...saw a mob
> pull a china house
> or rather up set whou it
> & the china woman boo hoo
> ing & old army coming
> along just at the moment
> the crash of the house & a yell[24]

While in 1855, Edwards documented Nelsonite actions towards the Chinese:

> A.M. The Miners in this vicinity having resolved that Chinese should not work on this portion of the Creek the following persons proceeded to eject 16 who were camped near Graveyard Flat they begged to be allowed to remain but remoustrance was useless as some of the miners commenced to pulling up their tent stakes upon which they packed up we escorted them to the Point saw them over the bridge at which the merchants drew long faces & gave us blue looks...
>
> Wrote notices calling a meeting of the miners on sunday 27th inst to pass special resolutions in regard to Chinese holding & working mining claims in this district (Nelson Creek)[25]
>
> Of all the miners in Nelson's Creek district including those who took part in the ejecting the Chinese and those who first complained of Williams & Teft selling their mining claims to the Chinese after having assisted and approved of the Chinese being ejected. Only the following persons appeared to respond to the call for a miners meeting to enact laws prohibiting Chinese working or holding mining claims in the district, E.A. White, Job Richards, Henry White, Smith, R Corson, W. Larrison John Young (alias Brigham) & Wm Edwards who hereby declares he declines taking an active part in any public business...[26]

A Chinese wedding held at Nelson Creek on November 25, 1865, was reported by the *Quincy Union*:

> A Chinese wedding was celebrated at Sterling Flat, yesterday. The ceremony was conducted in American style, and attended by a number of that "genus homo" from the Point, who thought it appropriate to kiss the bride. The last part of the ceremony must have been interesting indeed. Oh, ye gods!- that chapter on kisses would be appropriate. When can the heart be moved with more rapturous emotions? Oh, how sweet, how tender must be a kiss from Celestial lips!- ugh![27]

And, in September of 1868, contempt for the Chinese was again displayed when a "party of white men made a raid on a Chinatown on Nelson Creek near Nelson Point." Three hundred dollars were stolen, after which the thieves stole one hundred dollars from another camp. A posse unsuccessfully pursued them.[28]

The Chinese population at Nelson Point dwindled as the gold ran out near the end of the century. Among the last to reside there was Mi Woo, who in 1883-84 owned a house adjacent to Quong Chung's store.[29]

Long before Ellis Island, San Francisco Bay was the entry point to America for emigrants from all over the world. The contributions of those foreigners is evident. For the California gold rush was not only an American experience, it was a worldwide phenomenon.

NOTES.

1 Shirley, *Letters* (Santa Barbara: 1970), pp. 109-10.
2 Ibid., pp. 142-43.
3 Edwards, *Diary* (Bancroft Library: unpublished), vol. 4, Nov. 22, 1853.
4 John Haun, *John Haun Diary* (Plumas County Museum Archives: unpublished), Nov. 6, 1854.
5 Edwards *Diary*, vol. 11, Oct. 11, 1855.
6 Holliday, *World Rushed* (New York: 1981), p. 452.
7 Mary Dunn, *The Chinese in Early California* (Plumas County Museum Archives: unpublished).
8 Nancy Aki, *Principle Chinatowns, Mining Camps in NE California 1851-1900* (Plumas County Museum Archives; unpublished).
9 John W. Wunder, "Chinese in Trouble: Criminal Law and Race on the Trans-Mississippi West Frontier," *Western Historical Quarterly* 17, no. 1 (January, 1986): 38.
10 Ibid., pp. 25-41.
11 John Haun *Diary*, Sept. 23-24, 1856.
12 Fariss and Smith *History*, pp. 200-01.
13 Ibid., p. 202.
14 *Quincy Union*, Spring, 1864.
15 Dunn *Chinese in Early California*, p. 3.
16 Edwards, *Diary* (Bancroft Library: unpublished), vol. 11, Nov. 12, 1855.
17 Leroy Post, interview held at his home, La Porte, California, Jly. 24, 1987.
18 Robert Cotter, interview held at his home, Quincy, California, Jly. 25, 1987.
19 Edwards *Diary*, vol. 10, July 18, 1855.
20 Ibid., vol. 6, May 29, 1854.
21 John Haun *Diary*, Jly. 8, 1855.
22 *Quincy Union* 13, no. 34 (Jn. 17, 1865).
23 Ibid. 5, no. 52 (Oct. 26, 1867): 3.
24 Haun *Diary*, Aug. 29, 1854.
25 Edwards *Diary*, vol. 10, May 23, 1855.
26 Ibid., May 27, 1855.
27 *Quincy Union* 4, no. 6 (Dec. 9, 1865): 4.
28 *Plumas National Bulletin*, Sept. 28, 1911.
29 Nancy Aki, *Principle Chinatowns*, p. 24.

CHAPTER VIII.

BLACK POPULATION.

The black population in Plumas County during the gold rush, particularly during the early years, was numerically insignificant. Therefore, black contribution to the region's gold rush experience was negligible with the exception of one man—Jim Beckwourth. First, however, let me document the few recorded mentions of blacks in the vicinity of Nelson Point. Edwards noted a "Negro Bar" located on the Middle Fork of the Feather River near Poplar Bar. And in 1855 he, "Visited a Negro performance at Independence Bar $1."[1] Black minstrels were mentioned by Haun to have performed in Quincy in 1858 as well. And Haun noted, "an old Negro who worked at his ranch."[2]

The most noteworthy black person of the region was the famous mulatto mountainman, Jim Beckwourth. Because he was a professional guide and traveled the region extensively and because his routes from Marysville to Reno passed through Quincy and La Porte, it is not unlikely that Beckwourth passed through Nelson Point from time to time. However, this has not been documented. Haun only mentioned seeing him once while on a visit to the American Valley.[3]

Beckwourth collaborated with Squire Bonner to publish his autobiography, *The Life and Adventures of James P. Beckwourth* in 1856. In it Beckwourth displayed his legendary ability to stretch the truth and in true mountainman fashion often to exaggerate the facts beyond recognition.

The Black experience during the gold rush in this region was limited and, consequently, very few records exist to document the extent of their participation. Perhaps, further research will shed additional light on this subject.

NOTES.

1 Edwards, *Diary* (Bancroft Library: unpublished), vol. 10, May 21, 1855.

2 Haun, *Diary* (Plumas County Museum Archives: unpublished), Mar. 4-5, 1857.

3 Haun *Diary*, Oct. 22, 1855.

CHAPTER IX.

THE FEAR OF INDIANS.

> *...I have discovered some emigrant wagons [to report about] ...and found that they had lately come through hostile Indian country and had fared rather roughly...and felt that...I could add particulars that would make the article much more interesting...I put this wagon through an Indian fight that to this day has no parallel in history.*
>
> —Mark Twain

The fear of Indians was first instilled into the miners long before they reached the mother lode. Exaggerated reports about savage Indian attacks were popular reading in the east. These reports generated among many travelers a hysterical fear of Indians. In 1849, the U.S. War Department offered to sell guns and ammunition at cost to California emigrants and many of those traveling overland routes armed themselves like army battalions.

However, violent encounters with Indians were rare. A sample of 150 diaries, journals, memoirs, and letters assembled from all major overland trails, 1830-1900 indicated:

113 recorded no difficulties with hostiles
22 encountered minor problems
15 experienced major difficulties, but not extensive loss of life[1]

Regardless of the fact that life threatening Indian encounters were rare, the goldseekers' deeprooted fear of Indians often continued when they reached the mines. Certainly, in some instances there was cause for concern, but when examining the reports of Indian encounters in the vicinity of Nelson Point one must consider the fears that were imprinted on the minds of the miners. The following are accounts of Indian activity in the vicinity of Nelson Point:

Edwards wrote in 1855:

> Reported the Indians friendly had no trouble with them at all, but suspected they were preparing for War with the Whites[2]

Aylett R. Cotton who was camped on the Middle Fork 10 miles downstream from Nelson Point in early 1850, wrote:

> MOVING TO THE MIDDLE FORK OF FEATHER RIVER
>
> We engaged nineteen Indians to assist in carrying our Articles up to the Bar promising them shirts and pants. We proceeded a few miles the first day and camped

nearby a Indian settlement. Our boys killed two dear which were very plentiful as they had come from the Mountains to get below the greater depths of snow

One of the deer we gave to the Chief of the Indian Settlement.

The Indians had a circular structure of considerable size, what might be termed an Assembly Hall, having a contracted entrance through which to pass in; an opening in the top of the structure for the smoke to escape; ground floor, in the center of which a number of Indians... I obtained a violin in California, I do not remember how nor where nor what I finally did with it, and I sometimes played on this when these Indian acquaintances were at our camp, and knowing this, when we stopped for the night at their settlement the Chief invited us to come into this structure to have me play on the violin and have a dance. We complied.[3]

After attending a "dance" at the settlement, Cotton's party continued towards Nelson Creek. Cotton gave the Indians the clothing that he had promised before the completion of the trip. To his pleasant surprise, the Indians fulfilled their end of the agreement by finishing the trip.[4]

On the other hand, the *Quincy Union* reported:

INDIAN FIGHT.—We are informed that a company of soldiers overtook a band of fifty or sixty Indians, in the Black Rock country, and killed every one but one. A few more such "Indian fights," and the noble red man will conclude that the best policy for him to pursue, is to quit murdering white men. Keep up the good fight, say we, until they come to that conclusion.[5]

Charles D. Ferguson described another early Indian encounter:

We started in company with three more, for it was reported that the Indians were troublesome and it was not safe to go [to Rich Bar] in small parties. Nelson's creek was the first stream to cross. We camped one night in the mountains above the north fork, on a piece of marshy ground, where there were patches of scrub. In the morning our horses were gone. King and I went in search of them. After a little bit we separated and took different routes. Not long after, I struck the trail of the horses making down towards camp I was crossing over a sandy place, when all at once I heard something "zit" past, close to my head. Soon another "zit." I did not have to think twice to make up my mind what it was. As it is sometimes expressed, I did not run, but did some pretty tall walking. I did not look back, but before I reached a little rise of the ground, or spur of a hill, five of those "zits" had surrounded at regular intervals in my ear. As I reached the rise, which was not more than thirty, I saw a large bowlder that had rolled down from a high level. I jumped to the top of it and suddenly turned around just in time to see nothing but a bunch of long grass move a little. I up gun and fired into the center of the bunch, and left for camp. The boys had heard my shooting, and inquired what I had been shooting at. I told them a bunch of grass. "Did I kill it?" I said I did not go to look. We talked about that bunch of grass until there seemed to be a mysterious facination in the subject, and all

> had a desire to inspect it a little closer, so we all went together, taking of course, our guns with us. Arriving at the spot we were all surprised to find an Indian stretched out at full length, fast asleep, as we supposed, but on closer inspection we found he had a little bullet hole through him just under his armpits, He was taking his last sleep... We staid there two or three weeks and prospected around in different canyons with a little success;but on the whole it was not a paying trip, and we soon got tired. Armstrong wanted to go back to Long's Bar to his butchering business, and the rest of us were quite willing to. So we departed, stopping on our way back to prospect on Nelson's creek for a week, but we met with nothing encouraging. We never went out without our guns or revolvers, as we were liable at any moment to meet the infernal redskins. One day I was out looking for the horses and came across some acorns, the largest I ever saw, and gathered a few specimens to show the boys in camp. As I had them in hand, intently looking at them, I was wholly oblivious of anyone near me, but as I raised my eyes to start, I saw two Indians directly facing me, within forty feet, with their bows drawn and the arrows just ready to fly...I thought of my mother, my home, every act of my childhood; everything I had done in my life up to that moment, flashed across my mind in rapidly successive installments. Instinctively and without reflection, caused by the startling realities of the moment, as I now suppose, I flirted, rather than threw, the acorns in their faces. The act was a surprise to them and caused them to wince, and their arrows, which were sent at the same instant, to miss their mark. Now was my opportunity, and I availed myself of it. Before they could put their hand over their shoulder and draw another arrow from the quiver, I had my revolver, and in an instant there were two more redskins ready for the tanyard, but whose hides were too badly damaged to command full price.[6]

Hambly offered a sympathetic point of view in 1859:

> ...about 33 Indains from the Pitt River country they had been troublesome to the whites, or the whites to them, but <u>an</u> Indain has no rights, oh no, the aborigines has not any rights, kill them; kill them, take their lands, prostitue their women starve their children, introduce whiskey & civilization. & kill them[7]

The genuine, if not at all justified, fear of Indians in Plumas County led to the formation of the "Plumas Rangers." An independent military company, led at times by Nelson Point's Robert I. Barnett, and later by Noble C. Cunningham, numbered as many as eighty-two rangers. However, the Indian outbreak for which they were formed never materialized and even in subsequent years they never saw action. Consequently, they became simply a parade and drill unit and finally disbanded in 1860.[8]

Indian encounters involving real danger were rare throughout the northern lode. However, many miners shared a very real fear of Indians. This fear, more often than the Indians themselves, disrupted their lives and the production of the mines.

NOTES.

1 Glenda Riley, "The Spectator of a Savage: Rumors and Alarmism on the Overland Trail," *Western Historical Quarterly*, 15 (Oct., 1984): 443.

2 Edwards, *Diary* (Bancroft Library: unpublished), vol. 10, Aug. 29, 1853.

3 Aylett R. Cotton, *Across the Plains to California in 1849 And After: An Autobiography* (Bancroft Library: unpublished), pp. 37-38.

4 Ibid., pp. 37-39.

5 *Quincy Union* 4, no. 6 (Dec. 9, 1865): 2.

6 Charles D. Ferguson, *The Experiences of a Forty-niner During a Third of a Century in the Gold Fields* (Chico: 1924), pp. 80-81.

7 Hambly, *Autobiography* (Bancroft Library: unpublished), December, 1859.

8 Fariss and Smith, *History* (San Francisco: 1882), pp. 282-83.

CHAPTER X.

WOMEN.

In those days miners would flock in crowds to catch a glimpse of that rare and blessed spectacle, a woman!

—Mark Twain

Few women took part in the gold rush experience. The 1850 census calculated that women constituted only 8 percent of California's population. While by 1860, the number had increased to 30 percent, only 2 percent of that population resided in mining districts and many of those were prostitutes. Simply stated, women were rare—particularly at Nelson Point. Only a handful of women actually lived in Nelson Point.

Those who resided in the northern mining camps surely were a special breed. Perhaps the most celebrated of them was Mrs. Louisa Amelia Knapp Smith Clapp—Dame Shirley. Mrs. Clapp lived in nearby Rich Bar (N. Fork Feather River) and during 1851-52 wrote twenty-three letters to her sister back home in Massachusetts. Collectively, they were published as *The Shirley Letters*. I encourage the reader to peruse them for a beautifully written woman's perspective on life at the mines.

The California miner generally left his wife and family behind to maintain the homestead. Wives often experienced the rush only through their husband's letters. However, many miners either returned home and convinced their wives to come back with them to California or they persuaded their wives to join them by the letters they sent home. On the other hand, some men never returned home from the camps. On occasion, one of the "California widows," after as many as five years receiving letters ending with promises of a return "before the next rainy season," and after reading about opportunities for women in California, elected to pursue their men in that state—with or without permission.[1]

Upon arriving at Nelson Point, most women ultimately depended on their husbands for survival. Were it not for their desire to be with their husbands they would not have chosen to be there at all. Mrs. Peck, for example, left Nelson Creek shortly after the death of her husband in 1854. Hambly noted:

> ...at this time a Mr & Mrs. Peck who lived with, & was cooking for Mandonald (often called M Donald) was taken very sick & was soon at the point of death, myself & wife helped to relieve Mrs. Pecks troubles as far as our conditions would allow, but he succumbed to his fates, he was; so rumor had it hen.peck- be it as it may there was not

> much sympathy spilled over him by the family- the family was measureably destitute of necessary for comfort and in April (6th/1854) I went from Nelson to the American Val. and took up a subscription amongst the miners for the family...[2]

Edwards recorded their subsequent departure, "...Widow Peck & family left Nelson Creek for the States"[3]

"Respectable" women in the camps primarily played roles as subordinate business partners and as pioneer house-keepers. Business roles were generally limited to store clerks or boarding-house operators. Most wives at Nelson Creek operated boarding houses and/or laundry services for the miners. Mrs. Duesler, Mrs. Hambly, Mrs. Gilson, Mrs. Fox, Mrs. Pearce, and Mrs. Dever all engaged in these enterprises. The money was good. Mrs. Duesler, for example, earned for her laundry services, "3 bits apiece."[4]

However, the greatest profits made by women at Nelson Point were made while staring at the ceiling. That is to say, Nelson Point was the regional center for prostitution. Prostitutes comprised a significant percentage of women in the early days. "Public women," "fancy ladies," or "soiled doves," sobriquets by which they were often referred, accounted for a third or more of the female population in the camp. Because prostitution was "taboo," little was written about it and few records were kept. Yet references were occasionally made, as in the *Haun Diary*:

> The dance is to come off in a Gambling & Drinking saloon & to supper in Louis & Roots tavern. The Ladies had some idea of not attending as there had been two Public woman had been gambling there some time back.[5]

And Haun described the night's activities at Fagan's Saloon as ending with, "...& making love to the lady as they call it."[6]

Prostitutes in the camps were often Chinese. In August of 1854 Haun and Edwards both noted that six Chinese prostitutes operated at Nelson Point, however, they departed shortly after arriving.[7]

Women were also commonly employed as dealers in the gambling houses. Haun observed a woman dealing "Causkinnett" in a Nelson Point billiard saloon and also a woman to "draw custon" in Fagan's Saloon.[8] The following day he noticed, "two woman one in each house dealing Caskinett & lots of men to bet at her bank."

Clearly, the opportunities for women in the gold camps to make their fortune did not compare with those available to men. However, they did exist. Two extreme examples were the woman who made $18,000 selling pies to the miners and the prostitute who claimed to have made $50,000 in 1849.[9]

No evidence suggests that women were actively engaged in mining at Nelson

Creek, however, women often shared the lure of gold. The fascination generated by gold was often used to entertain women during social visits. Miners who might have shot other men for claim jumping, often encouraged ladies to engage in a sociable mining effort at their diggings. It was not unusual for miners to shower visiting ladies with gifts of gold-nuggets. Several episodes recorded by Edwards illustrate this point:

> Mrs Fox & Mrs Jane Edwards accompanied by Mr Hambly & W Prideaux & W.H. Blosser dined and spent the P.M. with us in a most convenial manner the ladies also visited the diggings & picked up $3. or $4. each on the bedrock...[10]

> Misses Swain Mrs Fox & Mrs Edwards arrived, the girls picked up and panned out $19.
> I also washed out a pan of dirt & bedrock for Madam's Fox & Edwards which with $9. worth of salt produced about an ounce
> After tea McGuire & I escorted the ladies home[11]

> Mrs Fox & Mrs Edwards escorted by McGuire & myself visited Haun's diggings and dined with him he gave the ladies some handsome specimans of gold I have never seen a man anywhere not even in California so much pleased or elated on the occasion of being visited by ladies[12]

Exactly how many women lived at Nelson Creek at any one time is impossible to say. During the early years of the rush approximately 10 "respectable" women and a half dozen or more prostitutes resided there. This was a mere fraction of the overall population at Nelson Creek. The scarcity of women occasionally created difficulties—the sort of problems that one might anticipate when too few women live in such proximity to so many lonely men. The Gilson-Wilson incident mentioned in the "Law and Order" chapter describes one such episode.

The social mores governing women were evident at Nelson Creek. Women were treated with respect and dignity. However, women's rights were only a subject for men to ponder, as Haun related:

> June 56
> Su 1 I lay about H.P. all day
> Gen Row came out & spent the
> whole day with us P.M. raind
> averry little, in the course of the
> day womans Rights were freely dis
> cusd.[13]

The precautions that were taken to ensure that women were treated properly is exemplified by women's customary use of the side-saddle on a return trip from Quincy to Nelson Point:

> ...arrived at the Illinois Rancho & in the absence of a ladies saddle fit for use Mrs White had to take a mans saddle James & Jane went ahead whilst we had to go on a walk on account of Mrs W. being unaccustomed to ride a man's saddle James & Jane having got several hundred yards ahead of Mrs W. become dissatisfied with being compelled to remain in the rear I shortened the stirrup on the off side & she threw her foot over & rode astride set of in canter soon overtook James & Jane who joined us in a laugh which made the forest ring arrived in sight of the Willow Rancho Mrs W. alighted I arranged her saddle she set her foot on my hand sprang into the saddle rode to the Point as upon a side saddle dismounted at Thompson's store & walked home[14]

The social scene at Nelson Point was occasionally blessed by the talent of a few of the Nelson Point women. It was not uncommon for the ladies to entertain the miners during the evenings at the Point:

> P.M. Attended a performance given by Miss Blodgett the Misses Graves & Mr Morton consisting of vocal & instumental music & dancing
>
> Of all the performances given in this vicinity except that of Miss Goodnow this was the most pleasing & entertaining the performance of Miss Blodgett on the Accordeon excellent Miss Graves's dancing the best I ever saw in fact I was so facinated I became either home sick or love smitten[15]

The most prominent woman to have resided at Nelson Point was most certainly Catherine Pauly. Mrs. Pauly operated the Pauly Hotel for nearly fifty years at Nelson Point. She became a beloved fixture at that location and may be considered the last official resident of Nelson Point having been forced to abandon the site only after the destruction of the hotel in 1924.

The few women who endured the hardships of the gold rush made lasting contributions. Although their role may not be considered significant in the early development of the gold camps, they played an essential part in the settlement of the region. The arrival of women often resulted in new long-term family plans. Many families found more than just gold in California. They found a future.

NOTES.

1 Holliday, *World Rushed* (New York: 1981), p. 453; Haun, *Diary* (Plumas County Museum Archives: unpublished), Dec. 4, 1854, Dec. 20, 1854, May 27, 1855.

2 Hambly, *Autobiography* (Bancroft Library: unpublished), April, 1854.

3 Edwards, *Diary* (Bancroft Library: unpublished), vol. 6, Apr. 22, 1854.

4 Haun *Diary*, July 22, 1855.

5 Ibid., Oct. 13, 1854.

6 Ibid., July 29, 1854.

[7] Ibid., Aug. 4, 1854.

[8] See the chapter on sport and leisure activities for a full discussion of the various gambling games.

[9] *Hunt's Merchant Magazine*, 1852; Holliday *World Rushed*, p. 303.

[10] Edwards *Diary*, vol. 11, Sept. 7, 1855.

[11] Ibid., Oct. 9, 1855.

[12] Ibid., Oct. 10, 1855.

[13] Haun *Diary*, Jn. 1, 1856.

[14] Edwards *Diary*, vol. 15, Jn. 17, 1857.

[15] Ibid., vol. 10, Jn. 8, 1855.

CHAPTER XI.

CHILDREN.

...what a rarity those embodiments of noise and disquiet are in the mountains.
—Dame Shirley

Few children lived in Plumas County during the early years of the rush. They were referred to as the "Infant Phenomena," indicating "what a rarity those embodiments of noise and disquiet are in the mountains."[1] Nelson Point appears to have been no exception to this general rule. The miners at Nelson Creek made very few references to children. Of the small number of families that were mentioned, it is unclear whether the offspring were adults or children. John Haun, who was the adult son of James, exemplifies this point. Consequently, the following families who lived at Nelson Creek during the early years may or may not have included children:

Dr. Vaughan family (1853)
Hambly family (1853)
Beach & family (1854)
Osbourne Madden family (1854)
Swain family (1855)
Richard Bass family (1855)
Collier family (1855)
W.B. White family (1857)
Sager family (1857)
Robinson family (1857)

Edwards offered the best estimate of the child population in 1856-7 in his accounts of a Christmas and a New Years party at Nelson Creek:

> Said [Christmas] party was composed of 5 ladies, 15 gents & 7 children[2]
> 7 ladies, 9 children & som 25 or 30 gents were present [at the New Years party][3]

Based on these figures, the mortality rate for children was staggering.[4] Hambly noted the birth of his daughter in 1853:

> ...my wife was confined of a little girl on 23 June right in the midst of all the troubles of sickness [small pox] &c...[5]

But less than six months later he noted:

> On the 3rd of January 1854 my little daughter Ophelia Anne Dee, passed to a better world, after a very short sickness. Doc Vaughan was called, but did not know her

> symptoms of sickness he was not a good Doc, nor thorough in his pharmacy, a gentlman when sober, but a sot, & tyranical, under liquor, he destroyed his share of "rot gut" and his wife and family suffered from his intemperate habits after he came to Quincy to live from Nelson Point.[6]

In 1855 a stillborn birth was documented:

> Dr. Hill informed us he had been called to attend Mrs Collier who was inomentary expectation of premature accouchment
>
> Mc Guyre gave up his bed to Dr Hill & took a share of mine
>
> About have past ten O,clock the Doctor came past the diggings and reported Mrs. Collier delivered of a Still Born Foetus
>
> A.M. Informed Mr. Collier that any or all of our Co were at his service not withstanding we were all Bachelors not posted up regarded the necessities the case called forth anything he wished done that we could accomplish not to be delicate about mentioning it and it should be promptly attended to whereupon he requested us to dig a grave & enter the Still born some 200 or 300 yards from the house agreable to which Mc Guyre & myself selected a spot dug the grave & entered the corpse myself acting as bearer & Mc followed in the rear in profound silence[7]

These were not the only infant funerals that Edwards attended, however:

> ...About 10 O,clock the infant son of W.B. & Elizabeth White died P.M. Bro James & H. Bray dug a grave
>
> I made a coffin. Edward A. White repaired the trail to the graveyard.
>
> Supper over, Edward & I lined the coffin with white glazed muslin & covered it with black carried it to the house & the corpse was placed therein
>
> Friday June 12th
>
> A.M. At 10, O,clock the following named persons assembled at W.B. Whites Cabin J.P. Prisk, W Prideaux J. Delevan, Bro James & wife & myself officiating as minister Prisk & Prideaux as bearers & proceeded to the graveyard where we intered the remains of Edward A. the son of W.B. & Elizabeth White who died June 11, 1857 aged 9 months I concluded the ceremony by reading the burial service of the Church of England...
>
> Evening I nearly completed engraving a head board for the Babe's grave[8]

The child population never increased significantly at Nelson Point because the terrain at Nelson Creek was too rugged for most families to seriously consider establishing permanent residence. Consequently, when the time came for a family to settle down those who chose to remain in the region most often set their roots in the American Valley.

NOTES.

1 Shirley, *Letters* (Santa Barbara: 1970), p.142.
2 Edwards, *Diary* (Bancroft Library: unpublished), vol. 14, Dec. 25, 1856.
3 Ibid., vol. 14, Jan. 1, 1857.
4 Infant mortality rates were often high on the frontier.
5 Hambly, *Autobiography* (Bancroft Library: unpublished), Jn. 23, 1853.
6 Ibid., Jan. 3, 1854.
7 Edwards *Diary*, vol. 10, Aug. 22, 1855.
8 Ibid., vol. 15, Jn. 11-12, 1857.

CHAPTER XII.

EDUCATION.

Very few families lived in Plumas County by 1852. They were still too scattered to make formal education practical. Because the few children living in the camps were called upon to perform chores necessary for their family's survival, even after schools did open children rarely attended past the age of fourteen.[1]

When Plumas County was organized in 1854 there were still no schools.[2] That same year, however, the 1854 June 9th Court of Sessions created three school districts, thus laying a foundation for a system of public education:

> District 1—Townships of Seneca Mineral
> District 2—Townships of Beckwourth, Quartz, and Goodwin
> District 3—Townships of Indian and Plumas[3]

A school fund of $141.93 was established for the county's 93 children.

For the school year ending October 31, 1855, the first superintendent of common schools, Dr. I.N. Hartzell, reported that

> District 1—Eighteen children lived within the district, but they were so scaterred a school was not possible.
> District 2—[Included Nelson Point, but was not mentioned.]
> Distirct 3—Seventy-five children between the age of 4 and 18 lived in the district. Twenty-nine could not attend schouol. The average attendance was 16.[4]

That District 2 was not mentioned indicates that the child population was insignificant and that no school existed.

By 1856, Plumas County's population began to shift. Elizabethtown (District 1) faded and Quincy became more prominent. Thus, in 1856 the Elizabethtown school closed and in October, the school districts were reorganized:

> District 1—Included all land in the county lying north of North Fork of the Feather River.
> District 2—Included all land between the North Fork and Middle Fork of the Feather River.
> District 3—Included all land south of the Middle Fork of the Feather River.[5]

Formal education moved one step closer to Nelson Point with the establishment of the Quincy School in 1857. Eighteen students enrolled, the average age being 8 2/3 years, with their ages ranging from 4-13 years old. Students utilized one text, the infamous *McGuffey's Reader.*[6] Existing records do not indicate that any Nelson

ABOVE:

Nelson Point School, Class of 1906

LEFT:

Mrs. N.E. Thomas

Point children attended. However given that Nelsonites regularly traveled to Quincy, an occasional schoolhouse lesson in Quincy was likely.

The School Census Returns for the year ending 1873, listed the enrollment of the Quincy School at 87 and La Porte's school at 104.[7] Nelson Point, again, was not listed, indicating the absence of a school. Those children from Nelson Point who attended school must have enrolled at either the Quincy or La Porte schools. In either case, regular attendance was not practical.

The Plumas County School Census of 1880 reported an enrollment of 984—an increase of nearly 1000% from 1854. The school fund had increased accordingly to $16,014.46—an average of $16.27 per child. A portion of these funds was utilized in the Nelson Point District which was formed during the May, 1880 term of the Board of Supervisors. The district boundaries were defined during the November, 1885 term and are outlined on Keddie's 1892 Map.[8]

The first teacher recorded for Nelson Point was Henry H. Childress in 1884. Henry was a native of California and twenty-six years old at the time. About 1890, a "good school building" was erected by a fund raised within a radius of a few miles of Nelson Point. The average attendance was 18.[9] The school appears on Keddie's 1892 Map, midway between Willow Creek and Middle Fork on the west side of the La Porte—Quincy Road.

Louis Myrtle was among the first Nelson Point school teachers.[10] And by 1905 Mrs. N.E. Thomas was the school teacher. Her enrollment was approximately nine.[11] In 1911, the schoolhouse was burned down by a 7-year-old student who didn't like school.[12] Portions of the Nelson Point District were annexed to the Pioneer and Quincy school districts during the March, 1921 term of the Board of Supervisors.[13]

A formal education was not easily obtained in Nelson Point. In the absence of schools, the family bore the responsibility for education. The quality of this education was limited to the amount of time afforded to home teaching, the learning materials available, and the educational level and teaching abilities of the parents. As time passed, schools became more accessible and the family's lessons were likely supplemented by an occasional Quincy schoolhouse lesson until children could be enrolled full-time in Nelson Point's own school.

NOTES.

1 Fariss and Smith, *History* (San Francisco: 1882), p. 313.
2 Ibid.
3 Ibid., p. 314.
4 Ibid.
5 Ibid.
6 Ibid.
7 *Plumas National* 7, no. 44 (July 26, 1873).
8 School file, Plumas County Museum Archives.
9 State Resources 11, no. 4 (Oct. and Nov., 1890).
10 Willow Ranch file, Egbert file, Plumas County Museum Archives.
11 School file, Plumas County Museum Archives.
12 Scott Lawson, interview held at the Plumas County Museum Archives, August 5, 1991.
13 School file, Plumas County Museum Archives.

CHAPTER XIII.

RELIGION.

We have no ministers, though fourteen miles from here there is...a man of distinguished appearance, an accomplished monte-dealer and horse-jockey, who is said to have been—in the States—a preacher of the Gospel.

—Dame Shirley

Religion did not play a significant role in the northern mining camps—particularly in the early years. On the frontier, organized religion was often slow to take hold due in part to the widely dispersed population. Even in Nelson Point where formal religion was introduced early, organized religion never enjoyed a popular following.

The first "church organization" in Plumas County was established in Nelson Point by J. P. Gentry as early as 1854.[1] Which doctrine he preached is not known, but when he moved to Taylorsville he spread the doctrine of the Methodist Episcopal (M.E.) Church. Haun did not mention Gentry even though he regularly noted the preachers of the region. Haun indicated instead that the American Valley was the center of religious activity. In Elizabethtown he noted:

July 54
Su 9 ...I then went to heare a
Methodist preacher the first
of any kind since Id been in the
country. & that too in my patchd
shirt & holes in my pants. No coat
or vest or handkerchief of any
kind & my old gum slippers &
mining hat and my butcher
knife at my side, but every
chap threw in his 50 cts when
the hat came around...[2]

Circuit ministers regularly preached in Quincy throughout the 1850's, but were sparsely attended.[3] In 1857, the apparent lack of religious activity prompted a visitor to note that Quincy displayed, "no clergyman, no Church organization, no religiuos services of any kind, nor is there any moral or social organization of any form,..."[4] Transient preaching continued in Quincy well into the mid 1870's: "Preaching Rev. Lewis Thompson will preach in the Town Hall next Sabbath at 11 o'clock AM and at 7:30 PM."[5] Haun failed to note that circuit preachers occasionally visited Nelson

Creek as well. Edwards cited a service one Sunday in 1855:

> ...I proceeded to Independance Bar to meet the Vigilance Committee, arrived there, a minister of the Gospel being present divine service was announced in the Bar room of the Sherwin House which was attended by 2 ladies (Miss Swain & Mrs Delavan) & about 40 gents during which the strictest order & decorum was observed the 11th Capter of Job was read & the 2 first verses chosen as the text. (Man that is born of a woman is of few days and full of trouble He cometh forth like a flower and is cut down: he fleeth also as a shadow, and continueth not) The transition of the mortal man was duly portrayed the absurdity of his accumulating riches which he could enjoy only temporarily & the actual necessaries of being one suit of raiment and one meal because if a man having a days provision and should die before noon he would have half a days provision more than was actually necessary etc This over...[6]

The following month Edwards noted another occasion:

> Suppered at the Sherwin Hous and attended divine worship in the bar room during which business was suspended the cards & chess board being removed to give place for the Bible & Hymn Book about 40 persons were present including Swain & family and the strictest order & decorum was observed whilst a short dissertation was delivered by Revd Bonner taking for his text the 10th & 13th verses (inclusive) of the 51st psalm and which was principly an explaination of the anatomical construction of the physical man with a few remarks on the conversion of St Paul and his faithful service of his Creation thereafter[7]

Hambly, on the other hand, demonstrated distaste for organized religion and it was reflected in his writing:[8]

> ...there was a Ball to be given at the American House to raise funds for the pseudo Methodist Church, raised over 100$ its as odd way to raise the wind for a minister & church Its a liberal move by the people, to build a house to be tongue lashed by bigots...[9]

> ...Dec Mo. 1860. this is this the 25 Inst, called Christmas day, to commemorate a farce of the Churches...[10]

The M. E. Church made the most significant attempt of any church to root itself in Plumas County. Early services were held by "circuit-riders" in "homes, halls, indoors and out, wherever circuit-riding preachers would get worshippers together." Services were held later at the Quincy Court House, but that privilege was discontinued by the county in 1872. Consequently, the women raised funds to construct Quincy's first church which was erected in 1877.[11] By 1882 the church had grown from its origins in Nelson Point to three parsonages. The churches and parsonages had an estimated value of $10,000. At the same time, no church ever numbered more than 75 members among the three circuits in the county combined.[12] Clearly, organized religion was unsuccessful in attracting a significant following at Nelson Creek or anywhere else in the region. Why?

Organized religion never played a significant role in the northern camps because of the very nature of miners. These men abandoned the restraints of civilized life in favor of a pioneer's existence. Most sociable miners preferred a saloon over a church. And even "God-fearing" men such as Haun, who regularly read his bible, seldom attended services. Nomen Nescio illustrated this point in a letter he wrote from Nelson Point in 1866. He said that men enjoyed Christmas not for its religious significance, but because it was a day for celebrating. "Shooting for chickens, was one way of enjoying it, and eating them, another..."[13]

Women introduced a significant religious influence to the gold camps. The ratio of women regularly attending religious services compared to the total number of women residing in the camp, far exceeded the same ratio for men. Dancing was popular during the 1850's, but by 1866, when a significant number of women appeared in Nelson Point, it was frowned upon. Strict religious mores became evident. Nomen Nescio wrote that religious service in the forenoon on Christmas was customary, but no Christmas ball was held because of the orthodox majority. "Dancing is scarcely thought of, and condemned `in toto' by all but four or five of the ladies."[14]

Perhaps, Rev. L. Ewing explained it best when he summarized the history of the M. E. Church in Plumas County in 1882: "This church has, true to its earliest zeal, pushed its ministers to the front of the county. And since 1854 it has steadily grown in strength; but the ever-shifting, changing character of the California miner is not to be depended upon for permanent membership in churches."[15]

NOTES.

1 Fariss and Smith, *History* (San Francisco: 1882), p. 315.

2 Haun, *Diary* (Plumas County Museum Archives: unpublished), Jly. 9, 1854.

3 Ibid., Jn. 22, 1856, Aug. 17, 1856, Jn. 27, 1858.

4 California State Agricultural Society, "Report of the State Agricultural Society," *Transactions: California State Fair and Exposition* (Sacramento: 1858-60), p. 207.

5 *Plumas National* 9, no. 50: 3.

6 Edwards, *Diary* (Bancroft Library: unpublished), vol. 8, Feb. 18, 1855.

7 Ibid., vol. 9, Mar. 18, 1855.

8 Hambly displayed opinions very antagonistic to organized religion throughout his writings — particularly regarding the hypocrisy and injustices of the Catholic Church and others. These beliefs, oddly enough, developed in later years into a strong belief in Spiritualism. In December of 1867 he noted, "...we had several callers...20 in all- after dinner 3 P.M. we had

quite a little social and in the evening had a Spiritual Seance, everything I believe was satisfactory to most of the parties- & some revelations, very strange through a stranger &c &c &c "Its not my province to say but little in this matter- Im simply an amature investigator, I do not charge any one, for, coming to investigate this strange feature of life, little known to man as yet, but Im going to look after it & hope ere long to be able to know & report more satisfactorly to all who may enquire after it, but I think the world is not yet quite ripe enough to understand its phenomena & facts-".

9 Hambly, *Autobiography* (Bancroft Library: unpublished), Jan. 1, 1859.

10 Ibid., December, 1860.

11 The Community United Methodist Church, *A Century of Service: 1875-1975* (1975), pp. 3, 9.

12 Fariss and Smith *History, p. 316.*

13 *Quincy Union* 4, no. 10: 1.

14 Ibid.

15 Fariss and Smith *History*, p. 316.

CHAPTER XIV.

LIFE AT THE MINES.

It is a kind of life that has a potent charm for all men, whether city or country bred.
—Mark Twain

Life in Nelson Point revolved around the mines. The hard work, daily drudgery, and solitude combined to produce a routine that was endured only for golden dreams of fortune. The Nelson Point lifestyle was not unlike that of other camps in the region, excepting its reputation as a center for vice. However, the gambling, drinking, and prostitution served only as distractions from a wearisome day at the mines. An accurate portrayal of life at Nelson Point comes only as a result of examining the details of day-to-day life.

A day in the life of a miner in 1850 was capsulated by Alonzo Delano:

> As soon as breakfast was dispatched, all hands were engaged in digging and washing gold in the banks, or in the bed of the stream. When evening came, large fires were built, around which the miners congregated, some engrossed with thoughts of home and friends, some to talk of new discoveries, and richer diggings somewhere else; or, sometimes a subject of debate was started, and the evening was whiled away in pleasant, and often instructive, discussion, while many, for whom this kind of recreation had not excitement enough, resorted to dealing monte, on a small scale, thus either exciting or keeping up the passion for play. Some weeks were passed in this way under the clear blue sky of the mountains, and many had made respectable piles. I highly enjoyed the wild life we were leading, for there were many accomplished and intelligent men; and a subject for amusement was or debate was rarely wantng. As for ceremony or dress, it gave us no trouble: we were all alike. Shaving was voted a bore; the air holes in our pants were not "few and far between," and our toes were as often out "prospecting" from the ends of our boots as any way, and two weeks before our supplies arrived I was barefoot, having completely worn out my shoes.[1]

HOUSING.

Frank Lecouvreur, who mined at Nelson Creek, described the typical shelter inhabited by the miners:

> Such things we soon learn here from one another; the whole life of a California miner is nothing else but a campaign. If he has a claim which gives him work for any length of time, he generally lives in a tent; drives four low posts into the ground, nails to these two pieces lengthways and two across; on these he nails a few potato sacks, and thus he has a solid bedstead. Only when he is traveling or prospecting he sleeps in the open.[2]

If a claim proved to be particularly profitable and if staying through the winter was considered, a log cabin was erected. The furnishings were mostly crude, but the cabin served its purpose. Haun described his residence:

Sept 1853
Su 18 ...After dinner we commensd to build a cabin.
We got up six logs 16 by 14 ft long. One other man to help.
Oct 1853
Tu 11 ...It rained all night. But our
bunk dose not leak, besides our
cabin is roofd chinkd & has a fire
place no door nor floor or window.
Sun 23 Tore down my bunk to get materials
for putting up one in my cabin, tonigh
I expect to sleep on a good sack
bottoms bunk in the cabin, & Dave &
John may sleep on my straw bed & on
the floor...
Tu 25 I slept easy but got cold under
2 pair of blankets last night. Got
up twice & made a fire. No gold today.
Dave & I went down to the point ...
I then bought 3 plates old at that
the first queens ware of any kind
that Ive used in months. Our furnitures
being tin plates old & black tin oyster
cans for fruit dishes & butter plates &
with wooden forks & spoons I am
now sitting on the first stool, whitch
places me above the ground or log of
wood that I made tonight since
midle of July past...[3]

CLOTHING.

A distinctive "style" of dress developed which proved to be practical for the northern lode. Kimball Webster described the miners' apparel:

> While in California [1850] I wore no coat. Shirts and pants were the clothing worn, and during the rainy season when it was cooler we wore extra flannel shirts. Blankets were much used instead of shirts when miners were not at work. At the time I was dressed with a "Hickory" shirt and thin pants, all cotton.[4]

Haun referred to his own attire:

Dec 1853
Su 4 Very pleasant in deed. We go
about with a cotton & flannel
shirt on, pants & no drawers,
& comfortable at that.

Sept 55
Mo 3 We got along verry well with our work, one of
troubles is to keep from getting wet, so we
on two gum coats as they are leaky...[5]

By the mid-50's India rubber apparel was widely worn by the miners. India rubber boots, for example, were worn to combat the cold waters of Nelson Creek. Rarely did water temperatures rise above 50 degrees, even during the warm summer months. Gum coats and gum overalls were common attire as well.

New clothing was often in short supply. Consequently, the mending of clothes became a regular chore:

July 1853
We 20 ...Tonight I put a whopping
big patch on the right knee
my blue military pants out of
twillo cotton flour sack.[6]

DIET.

When convenient and affordable the miners took meals at taverns at Henpeck Flat, Independence Bar, or at the "point." However, most meals were consumed at camp. The availability of some foods occasionally limited the diet of miners. However, given the remote location, a surprising variety of food was consumed. The recent development of canned foods contributed to making this possible.

Three meals were generally consumed each day, although often each meal consisted of the same items. Haun regularly described their diet:

June 1853
Sat 25 Our breakfast was of the same,
that ours supper, consultation...Took
dinner and supper of the same,[7]
stopped at the inn eat Breckfast
each of us $

Sept 1853
Sun 18 ...What a gratification it is
to have an appetite to suit the
times. We came back to camp
& took dinner good light bread & tea
beef stake potatoes & onions boiled
beans & butter.
June 1853
Fr. 24 ...We came down again to N. Creek
& took tea. Our supper consisted of
soda crackers cold roast beef & midlen
meet & water...
Sept 1853
Sun 4 ...For dinner we had dride apple pie
for desert.
June 54
Th 8 ...I am buying milk at
6 Bits a gl.
Augst 1853
Th 19 Eat pork & beans in consquence
of whitch dreamed that I was
in george town...
Sat 20 We finished digging our ditch &
let the water in, had supper
tea bacon & bread...
Sun 21 All hands at camp sitting around
a water mellon that cost 5$.
Ive had as good for 25 cts.[8]

LONELINESS.

Solitude was intrinsic to the mountains, and so, an overwhelming theme among gold rush camps was loneliness. Miners relied on a variety of diversions to take their minds off of their families, homes, and friends. They frequented saloons and bowling alleys, gathered around campfires, wrote letters home, read novels—all in an effort to avoid becoming overcome by loneliness. Haun's writings very clearly illustrate his loneliness:

July 55
Th 26 ...I am all a lone of the once
great mountain company making these
little entrys, not a voice to be
heard except the screeching of a grass

hopper, the wind blowing sotly throug
the tops of these tall pines

Augst 55
Su 12 ...I shed many tears of grief
while thinking of our lonly condition
& that of my poor wife seperated
as we were. I tride to sing
some old fashiond hyms, but I
am in a strage land, my harp
is hung upon the ceaders...[9]

Friendships played a very important role in fighting off the desolation. Bonding between friends often became very strong and, consequently, when friends parted the goodbyes became difficult:

Sept 55
Fr 28 I dreamd of wife again. I cer
tainly will see her soon. I calld up
all hands at day light, Rositer had
got back. After breckfast we settled
up & pd 8$ for beef & tea, our shear
& then weighd out in gold dust
at 17 1/2$ per oz 100$ to loan Rains
to pay Rositer. He divided out his
wearing apparell to us & bid us a
final a dieu. It braught tears in his
eyes to leave us. I thruo an old shoe
after him. We then went to work
with the pipe till noon...[10]

Separation from their wives produced severe feelings of desolation for some miners. Letters received from home did much to alleviate the anguish, as did the therapy of writing a letter home. Haun became extremely lonely for his wife, reflecting this in his writing:

Oct 1853
Su 30 Got up before day built a fire
lit a candle drew up my center
table & got down my carpet wallet
& took my bundle of letters
& read them over carfully & the
last one I got from you is
of date June 27th 53...

June 1855
Sat 9 ...I feel lonesome & disappointed on account of my wife not coming out.

Apr 1854
We 5 ...I bought his bed, four pair blankets straw bed & sack bottom all for 1/2 oz gold & slepd in it with old Goodshall along side of me. I would have pre fird my wife.[11]

SEX IN THE GOLD CAMPS.

The relative absence of women in the camps presents an interesting case study for psychologists. This subject is rarely discussed, however, and it is appropriate that some mention of it be made. That so many men existed with so few women for so many years is remarkable. How did they do it?

The first and most obvious answer was prostitution. Alcohol and prostitutes were among the first "supplies" to arrive in a new gold camp. Certainly, Nelson Point offered more than its share of prostitutes. For a variety of reasons, however, many men did not satisfy themselves in this fashion. Instead, they were often psychologically tormented by the lack of female companionship. This became a frequent theme with Haun. In some instances he simply longed for the presence of women, in other cases his writings took on more sexual themes:

Nov 54
Th 16 I dreamd Last kt of asking Susan Wallace if she would not marry, I thought she told me there was two many pains & achs in a married life. I then dreamed of kissing another women that I had met in my travels I for got who she was...

May 55
We 23 ...I dreamd last kt of being quite feelingly situated with ann EW—.

July 55
We 19 ...I dreamed that Jane Cooper

was a widow & had 2 children
& that she wanted me to marry her.

Augst 55
Th 16 ...I dreamed of wife the
last two kts in connection
with some other young woman
the first with Lizzy & last kt.
Miss Jes Cannon.

Sept 55
We 12 I dreamd of wife having been
travling on the sea & that we were
together. I thought Jane Cooper
was with us & that she was in
gerat distress but quiet I
thought I kissed her & wife
said she was so sorry for her.
I did not think that Jane was
married...

Sept 1853
Su 11 ...I dreamd that I was in
Georgetown & that you wanted to
to bed with you.....——-

Aple 55
We 11 ...I dreamd Mon. kt last that
I was in company with my wife
& she denide me the gratification
that is common with man & wife.

May 55
We 30 I dreamd last kt that I was
in bed with wife & I thought
she rather refused to indulge
in that sotial intercours as
man & wife well it is no go.

Sept 55
Su 9 ...I dreamd of being in
bed with wife & thought I
was enjoying my self & felt
pleasant twice & wakd up
making the third effort, but
all a dream.[12]

Among the most intriguing of Haun's dreams involved the theme of prostitution:

Sept 55
We 26 I dreamd last kt of being in
company with wife again. I thought
that we were together in some
city alone, she had on a dress
for bed I thought. I asked her
where she slepd, she told me
in some tavern, but I forgot
the name of the house. I then
asked her if she could accomodate
me, she told me if I would
give her all the money I could
conveniently do with out,
that she would. I was
about to sware by God
& I changed to the holy
St. Patrick. I was so
mutch vexd that I
woke up immediately & did
not sleep good again the
last half the kt. Besides
Ive felt uncomfotable
all day & the dream in my
head all the time.[13]

No evidence exists to suggest that homosexuality was practiced in the camps. If it was prevalent, it was not openly discussed.

ALCOHOL.

By far, the most popular diversion indulged in by the miners was drinking. An enormous quantity of alcohol was consumed throughout the mother lode. In 1853 California imported 20,000 barrels of whiskey; 400 barrels of rum; 9,000 casks, hogsheads and pipes, 13,000 barrels and 23,000 cases and boxes of beer; and 5,000 pipes and casks, 6,000 barrels, 5,000 kegs, 8,000 cases and 1,600 packages of "unspecified liquors."[14] Nelson Point certainly drank its share. In addition to frequenting Nelson Point's two saloons, the miners drank in camp, around campfires, alone, at dances—and very often in excess. Intoxication was commonplace and undoubtedly it affected production at the mines. Alcohol became an integral part of life by acting as another means of staving off loneliness at Nelson Point.

A HAZARDOUS LIFESTYLE.

Beyond the dangers inherent in mining, the rugged lifestyle in the Sierra Nevada exposed Nelsonites to other significant hazards. Accidents occurred frequently. Often they were fatal. Simply walking along the creek sometimes resulted in mishaps. For example, Edwards noted two incidents in the spring of 1854:

> ...Evening Spencer fell in a Cayote Hole about 16ft deep cut his head which bled profusely[15]
> ...[Edwards] went to Nelsonville in returning slipped of a ledge of rocks and lacerated one of my fingers badly.[16]

Haun described other accidents in which routine tasks produced tragic results:

> Sept. 1853
> Mo 19 ...After dinner Dave
> was at work knotching down
> alog on the cabin & stuck
> the ax in the instep of his
> left leg he sewd up the cut
> himself...[17]

Dave's cut later infected and caused a serious fever. Haun also related an accident involving Frank Walker:

> Augst 55
> Sat 25 A man by name of Frank Walker went
> up to the bridge that cosses
> feather River above Nelson Creek
> stripd off to take a wash
> & made a dive struck his head
> against a rock, those present
> became alarmd on account of his
> long absence under water & then
> seeing the blood in the water
> next saw him strugling, they sprang
> in & hawld him out. That was on
> Friday evening 24th & this
> evening he is dead...[18]

In 1864 Mr. S. Allen broke his leg while working on his claim. Two weeks later Dr. Cate and Dr. Vaughn were forced to amputate it just below the knee.[19]

Parker recorded one of the earliest fatalities on Nelson Creek:

March 30th [1851] Nelson Creek.
...A sad accident has today happened a mile below here. Some miners were cutting a large tree that threatened to fall on their camp. As the tree fell, it struck against a tall cedar and bent it down some distance before it became disengaged. The cedar sprang back and snapped off, falling the other way from which it had been bent. A club struck one man on the head and broke his skull, killing him instantly. The tree broke another man's leg, and hurt still another slightly.[20]

Hambly reported the death of Limon Arnold:

...-On the 6th of January 1853 Limon Arnold was killed by a rock striking him in the forehead as he was returning from work. I was deputized to sell his diggins & send the same to his wife at Prattville Wisconsin.[21]

And Edwards described the details involving the death of James Trumble:

...James Trumble fell from a log in the act of crossing the creek and was drowned[22]

The following month:

...Sterling found the body of the late James Trumble in a pile of driftwood opposite Meaker's Flatt an inquest was held on the body the verdict was; came to his death by accidental drowning, many persons disputed its being the body of Trumble until when stripped his name in full was found tattooed on his left arm, born 1821...had been a sailor said to be from the vicinity of Rochester NY[23]

The primitive conditions under which the miners lived and the lack of proper medical care contributed to the spread of illness and disease at Nelson Creek. In their diaries Haun, Edwards, and Hambly all reported that the miners were constantly overcome by sickness. Seldom did they offer details regarding the symptoms, rather, they made clear that a great deal of time was spent convalescing in camp. The most serious illness recorded was the outbreak of small pox at Nelson Creek in 1854, which necessitated opening a hospital. When one of the doctors was overcome with the disease and the other refused to treat the ill, Hambly remained to operate the hospital:

Jan. 53
...I was a whip of the very liberal type altho in the small pox cases that broke out on the Creek I took care of them to the last. Edmund Bridger & one of my children the only ones that succumed to the disease, altho Doc Hill with several others was down with it at one, time, and Doc Vaughn forsook the sick and helpless.- I staid with them to the last and was voted by Doc Vaughn this clique, "serve him right" and that was my pay- Mr Cowney, Cap. of the Wisconsin Comp. paid me for waiting on Bridger. he Bridger. died from cold & exposure driving the pox internally, before Vaughn, or Hill would

acknowledge he had the disease, they pronouncing it the measels, after he was compelled to remain in a cabin got for the purpose- he lived only a week, we buried him in a shaft close by the Cabin & burnt the Cabin, he had been boarding with Fred. Mandonald on Hargrave Flat subsequently call "Henpecked" Flat there is a signal to this name. X and mixing very freely with the miners & it was very fortunate we got it no worse in the Creek only 7 Cases in all.[24]

THE DAILY ROUTINE.

Life at Nelson Point was not entirely consumed by the process of mining. A significant amount of time was spent addressing the details of day-to-day existence. Haun's Diary offers the most comprehensive view of life at Nelson Point:

Oct 1853
Mo 17 Took out today $16=1/2
Tu 18 " " " 12 1/2
Dave trimd my hair by candle
light. John baked 3 loves of
bread & then wasd shirts towels &
socks and I wasd & went to bed
the kiotes gives us a barking evry night.
Nov 1853
We 16 Commensd snowing early in
the morning. It son stopd at 4
inches deep, three of us went up
the ditch hunting. By knight
the snow was quite gone, cut
wood & lay it in the corner to
have it
handy as I get up from 1 to 3
times in the knight. I keep a good
fire all knight. John dose the cooking,
have not dug since Friday last.
Th 17 Beautiful day 7 hands went to
work got gold today $14 1/2. My
Virginia friend is gone this ev
ening over to the American Vally
to take up claims for us three,
if he gets a good prospect.
John has just washd out 1/2 doz
flour sack to line his pants with
& now lying in bed reading some novel.
Frf 18 Clowdy all day. Got gold 22$

I made a new patten riffle box
to katch gold. 2 hands set up one
set of boxes near our cabins.

Jan 1854
We 25 before day it was snowing & con
tinued until about 10 am & then
cleared off warm & pleasant. We went
over to see if there was water to work, but
there was not enough. We waded back to
the cabin through the snow. It is moore
than knee deep & occationally it holds
us up. We cut & split wood & pild up un
der the shed along sid of the house the
rest of the day. We are sitting around
the fire. Dobson & John reading novels
& I making this entry. All is still &
quiet except small bits of frozen snow
that falls from the trees on the cabins roof.
I think of you wife allways when I
write in this book. It seems to me that
I can see you reading it & asking questions
& saying I would have done thus & so.

July 54
Mo 31 Shaw & I got 19$ &
John, Cook the man th
at works for Lloyd got no
gold to day. Lloyds finger
is getting well fast.
I bathe every kt.
raining about sun set, though
but light. I & John are sitting
in the cabin taking our ease. He
is trying to play a new tune
on his fiddle & Ive wrote this down.[25]

In later years, the best source of information about the lives of Nelsonites became the "Letter From Nelson Point," published regularly in the *Quincy Union.* The most faithful correspondent was Nomen Nescio. His letters offer a unique view of the times and appear in Appendix IV.

Nelsonites were a special breed that characterized the pioneer spirit of the times. But even strong men became discouraged. Adolphus Windeler wrote:

> Sept 3, 1850
>
> Friday 28, turned out early & started for Nelsons Creek, which we reached about dinnertime, along steep trail leads down to it. While sitting in our camp we heard something come down (tumbling the hill. Went to see what it was & found a lot of pork etc & a mexican packsaddle. Pick it up & put it so that the river could not carry it off. They owners came soon after & told us they had lost two mules & three packs already. Slept in the sand without Shelter of fire. Pretty comfortable for all that, for it is warm in the valleys, whilst it is cold in the mountains...
>
> Sund. 25, washed cloths, looked around on Nelsons creek, bought 4 ld potatoes...Wednsd. 28 fixed a bushhouse. Here we be, live on flour Apples & pork, work like horses, making our living, sleep sweet & got no trouble...So we live on, don't know yet how to work during winter who is coming along. Mice so thick that they even bite us. Bake bread at night for the next morning, & boil dumplings enough for dinner to last us at supper...I am plagued with sores on my arms. Lived all this week on flour & water with some dried apples...My left arm being no better went to Dr Vaughan who told me it was scurvy...I am pretty near giving up hopes of making a descent pile of money to go home with. But then I may as well stick it out here some years as not for to this winter, it is out of the question to make money enough to go home with...[26]

For many the hardships proved too much. Hambly noted:

> My Brother William getting very dissatisfied with such a rough life left the Creek on the 9th of June 1853 for Australia by way of San Francisco...[27]

Clearly, for most it was a rough life at Nelson Point. For others an adventure. The days were long at Nelson Creek and the conditions were harsh. And Nelsonites subjected themselves to working and living conditions that were not only primitive, but often extremely dangerous. When they weren't breaking their backs shoveling a sluice they were often convalescing from an injury or illness. Or they were busy mending pipes or a pair of trousers. They were isolated, often lonely and lacked female companionship. Why did these goldseekers subject themselves to such a lifestyle?

One could argue that the wages were very high. The six to eight dollars a day earned by most laborers at a time when cowboys were earning a dollar a day working cattle, may have been reason enough. But, goldseekers did not venture to California to become laborers. They only resorted to day labor when their own pursuits had failed. Those who remained in the face of so many who were returning home, did so in part for another reason. They were afflicted with a condition more addicting than any opiate—gold fever!

Gold fever is difficult to describe to those who have not experienced it. But simply stated, it is a state of mind that can induce a man to overcome any obstacle often with careless disregard for his own well being. It can motivate a man to work himself into a state of such fatigue that he barely has the strength at night to chew his supper. Yet, each night he dreams of the fortune that awaits him the next morning at bedrock. That's what kept them going.

NOTES.

1 Alonzo Delano, *Across the Plains and Among the Diggings* (New York: 1936), p. 153.
2 Frank Lecouvreur, *From East Prussia to the Golden Gate (New York: 1906), p. 244.*
3 Haun, *Diary* (Plumas County Museum Archives: unpublished), Sept. 18, 1853, Oct. 11, 23, 25, 1853.
4 Kimball Webster, *The Goldseekers of '49* (Manchester, N.H.: 1917), p. 162.
5 Haun *Diary*, Dec. 4, 1853, Sept. 3, 1855.
6 Ibid., July 20, 1853.
7 Haun described traditional mealtime nomenclature. Breakfast was the same as we know it today. However, what he referred to as dinner was taken at the time we eat lunch today. And what he referred to as supper was consumed at the time we have dinner.
8 Ibid.
9 Ibid.
10 Ibid.
11 Ibid.
12 Ibid.
13 Ibid.
14 Holliday, *World Rushed* (New York: 1981), p. 454.
15 Edwards, *Diary* (Bancroft Library: unpublished), vol. 6, Apr. 17, 1854.
16 Ibid., May 21, 1854.
17 Haun *Diary*, Sept. 19, 1853.
18 Ibid.
19 *Quincy Union* 2, no. 46 (Sept. 10, 1864): 3.
20 Dr. William Tell Parker, *Notes By the Way: a journal of the gold-rush days* (Huntington Library: unpublished), p. 145.
21 Hambly, *Autobiography* (Bancroft Library: unpublished), January, 1853.
22 Edwards *Diary*, vol. 6, May 5, 1854.
23 Ibid., Jn. 14, 1854.
24 Hambly *Autobiography*, January, 1853. [It appears that Hambly was off by a year on his recollection of this event. This took place in 1854].
25 Haun *Diary*.
26 Adolphus Windeler, *The California Gold Rush Diary of a German Sailor* (Berkeley: 1969), pp. 92-94.
27 Hambly *Autobiography*, June, 1853.

CHAPTER XV.

SPORTS AND LEISURE ACTIVITIES.

Swearing, drinking, and card-playing were the order of the day, and occasionally a fight was thrown in for variety.

—Mark Twain

Since the dawn of civilization sport and leisure activities have been integral aspects of life. Although the interrelationships between sport-leisure activities and culture continues to be argued by sociologists, this much is indisputable: sport and leisure activities act as diversions in mankind's daily struggle for existence. It should come as no surprise that in gold camps, where hard physical labor was a way of life, miners sought escape. They found it in a wide array of activities, ranging from sports to the theatre.

SKIING.

John A. Thompson arrived in California in 1851. He quickly became familiar with the difficulties the heavy snows created for the northern miners. With the aid of a little valley oak he developed the earliest of solutions—a pair of skis. A native of Norway, Thompson easily adapted his childhood skills to the Sierra Nevada. His 1856 return trip from Genoa, Carson Valley, to Placerville in just three and a half days, instantly established skiing (snowshoeing, as it was then called) as a northern lode institution and "Snowshoe" Thompson as the father of American skiing.[1]

Thompson's Norwegian snowshoes changed the miners' outlook on winters in the northern camps. For the first time, snow became a source of true recreation. Men skied from camp to camp solely for pleasure.[2] Skiing became a means of getting out of a cold and dark cabin and enjoying the often idle times at the mines during the winter.

Skiing was introduced to Plumas and Sierra counties in 1858 and by 1859 it had "become very common." The skiing "mania" spread like an epidemic throughout Plumas County. By 1863, mining towns scheduled races nearly every week of the ski season and many boasted of champion snowshoers. Entire families competed when races were held for both women and children.[3]

Nelson Point was located on the fringe of ski country and, consequently, it is not likely that races were staged there. However, races certainly occurred nearby. Contestants were known to have come from La Porte, Sawpit Flat, Onion Valley and nearby

Early snowshoe race, La Porte

Poorman's Creek.[4] The evidence suggests that Nelsonites competed as well. By 1867, La Porte emerged as the center of Sierra skiing.[5] That "snoe shoes" were mentioned in a Nelson Point theatre production suggests that skiing literally took the northern lode by storm.[6]

BOWLING.

Bowling alleys were prevalent throughout Plumas County during the gold rush. By 1855, ten bowling alleys were operating in the county and at least one of these was located at Nelson Point. By 1854, Bath, Powers, & Co operated a bowling alley in Nelson Point.[7] And, in 1855, Edwards noted:

> The new Ball Alley & Saloon being opened all were invited to enter and amuse themselves in which E.A.W. & myself participated[8]

Based on Edward's numerous accounts, bowling appeared to be among the two most prominent sports in the region.

BASEBALL.

Baseball was introduced into the northern camps during the 1860's and soon

Snowshoe race spectators, La Porte, 1902

enjoyed great popularity. The organization of Plumas County's first team came in 1867:

> Base Ball- Arrangements are now being made for the formation of a Base Ball Club in this town. The rules governing the game, together with instructions for forming clubs, have been sent for, and when received will be organized. We have no hesitation in saying that we believe a club can be organized in this place which, with a little practice, would compare favorably with any in the State. We hope some of our sister towns will also organize clubs, and when the proper time comes send the Quincy boys a challenge. By all means let's have the "National Game" introduced into this county.[9]

Baseball enjoyed tremendous favor in the years that followed. By 1887, a

Courtesy Plumas Eureka State Park Association

Snowshoe race announcement

Quincy Colts Baseball Team, 1906

sufficient number of teams had been organized to support a local tournament. Two of the local ball clubs that competed were the Roosters and the Hornets.[10]

It is not clear that Nelson Point ever fielded a team. And given the lack of level land for a practice field and because the town was already in significant decline by 1867 when baseball was organizing in the county, it is not likely that one was ever organized. It is probable, however, that Nelsonites competed on teams in the American Valley.

Because the local press seldom covered sporting events, little is known about the competition. However, the existence of a variety of team and individual photographs suggests that baseball's popularity in the region continued to increase through the turn of the century.

HUNTING–SHOOTING–FISHING.

To what degree hunting was a necessity rather than recreation remains unclear. Deer, elk, and bear were the main sources of meat in the early years. Even by the mid

1850's, when livestock was raised locally, hunting wild game often supplemented the miners' diets. Nevertheless, at times hunting was considered as recreation by the miners. While staying over at the New England Ranch, Edwards "Spent the day about the house joining in the various games with which the boys amused themselves...hunting Squirrels.)"[11] During another outing at the Illinois Ranch Edwards recounted:

> Monday June 5th
>
> Went hunting A.M. saw nothing worth shot & powder P.M. crossed the ranch to Martin's killed 2 doves wounded a grey squirrel which out ran me & shot a grouse twice & that flew away. shot a large snake which appeared too lazy to get out of the trail thus ended the day of hunting[12]

Target shooting was a favorite among the miners and was regularly noted by Edwards. On special occasions Nelsonites partook of an activity called "shooting for chickens."[13] It is not clear whether the chickens were the targets or simply the reward for good shooting.

As might be expected, fishing was a favored past time. Edwards, for example, mentioned his outings on many occasions. Not only did fishing offer recreation, but, like hunting, it supplied food as well.

HORSE RACING.

The sport of kings was formally introduced to the northern lode with the organization ofThe La Porte Jockey Club in 1867 and the Quincy Jockey Club the following year.[14] Already avid race enthusiasts, Nelsonites were instrumental in organizing the Quincy club. Three of the five original club vice-presidents were Nelsonites: J.T. Byers, J. Hargrave, and S.T. Brewster.[15]

Both trotting (harness) and running races were held as late as 1890 at Rockwell Park in Quincy. Rockwell Park was built by the Quincy Jockey Club and Col. Calvin W. Rockwell. Running races ranged in length from 1/2 to 1/4 miles. Other equestrian events were staged there, such as "Ladies Tournaments." In this event which included loping, running, and jumping their mounts over hurdles the women competed for cash prizes.[16]

However, horse racing did not begin with Jockey Club activities. On the contrary, racing activities pre-date these Jockey Club activities by more than twenty years. Haun and Hambly were both race enthusiasts. Haun made references to the horses and recorded races in the region as early as 1855:

Nov 55
Sat 17 ...We got across
Spanish creek than runs throu
gh the valley. We stopd to a horse
race & saw two...[17]

By 1858 horseracing was already well organized:

July 58
Tu 6 ...the races commensd to day
5 entries stinsons Woodpecker
winning the purse
We 7 the same horse won the race
to day. I was not out on the track
to day but was yesterday
Th 8 five horses started today
wheelers pided mar won the
race...[18]

Woodpecker was among the finest horses of the day. Hambly noted another Woodpecker start the following year, "...racing this week, high, times with the Jockying. Woodpecker the winning horse..."[19] In September of the same year Hambly wrote. "19.20.21. racing on the Plumas course: Billy won two purses, on the horses he rode..."[20]

There was little feed at Nelson Point and the area was much better suited for mules than horses. Consequently, Thompson kept mules and pack horses, rather than racing horses, in his stable at the Point. Nonetheless, racing enthusiasts abounded there, not surprisingly given the combination of gambling's popularity with the frontiersman's legendary love of horses.

WRESTLING.

Informal wrestling matches were very common on the frontier during the 19th century. Abraham Lincoln, for example, was among the many backwoodsmen and frontiersmen who enjoyed the sport. Most matches were wrestled from the feet, the object being to throw one's opponent to the ground.

Little is known about informal matches in the gold camps. However, on occasion, formal contests were organized. These contests were "all-comers" events, lasting several days, and drawing contestants from all over the county. English and American style wrestling were popular and often cash prizes were offered.

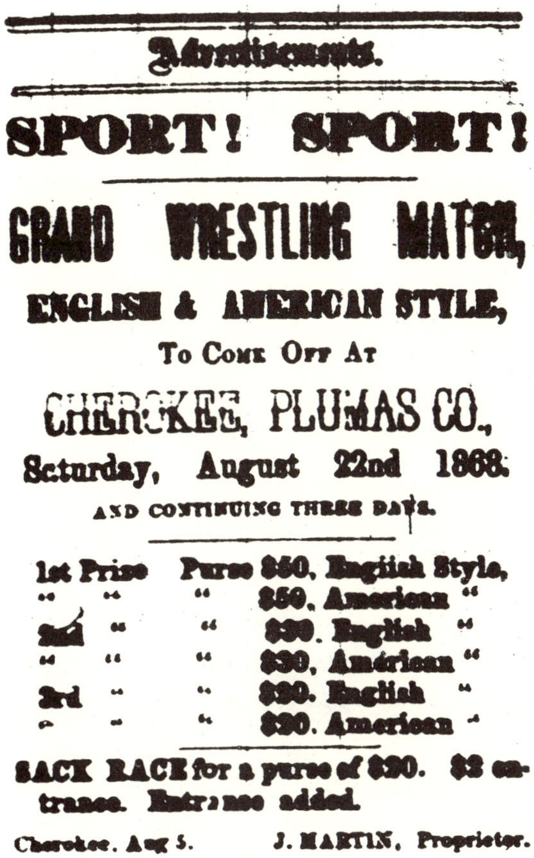

Advertisements.

SPORT! SPORT!

GRAND WRESTLING MATCH,

ENGLISH & AMERICAN STYLE,

To Come Off At

CHEROKEE, PLUMAS CO.,

Saturday, August 22nd 1868.

AND CONTINUING THREE DAYS.

1st Prize	Purse	$50, English Style,
"	"	$50, American "
2nd "	"	$30, English "
"	"	$30, American "
3rd "	"	$20, English "
"	"	$20, American "

SACK RACE for a purse of $20. $2 entrance. Entrance added.

Cherokee, Aug 5. J. MARTIN, Proprietor.

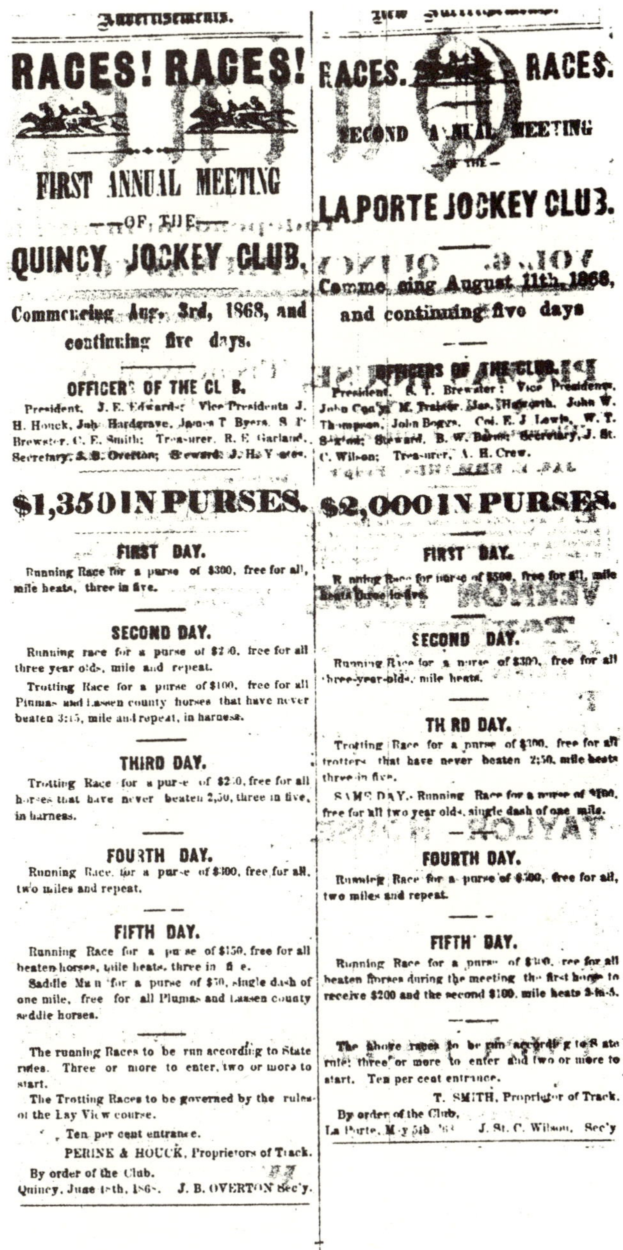

Advertisements.

RACES! RACES!

FIRST ANNUAL MEETING

OF THE

QUINCY JOCKEY CLUB,

Commencing Aug. 3rd, 1868, and continuing five days.

OFFICERS OF THE CLUB.

President, J. E. Edwards; Vice Presidents J. H. Houck, John Hardgrave, James T Byers, S. P. Brewster, C. E. Smith; Treasurer, R. E. Garland, Secretary, J. B. Overton; Steward, J. H. Yates.

$1,350 IN PURSES.

FIRST DAY.

Running Race for a purse of $300, free for all, mile heats, three in five.

SECOND DAY.

Running race for a purse of $2[illegible]0, free for all three year olds, mile and repeat.

Trotting Race for a purse of $100, free for all Plumas and Lassen county horses that have never beaten 3:15, mile and repeat, in harness.

THIRD DAY.

Trotting Race for a purse of $2[illegible]0, free for all horses that have never beaten 2,50, three in five, in harness.

FOURTH DAY.

Running Race for a purse of $300, free for all, two miles and repeat.

FIFTH DAY.

Running Race for a purse of $150, free for all beaten horses, mile heats, three in five.

Saddle Race for a purse of $50, single dash of one mile, free for all Plumas and Lassen county saddle horses.

The running Races to be run according to State rules. Three or more to enter, two or more to start.

The Trotting Races to be governed by the rules of the Bay View course.

Ten per cent entrance.

PERINE & HOUCK, Proprietors of Track.

By order of the Club.

Quincy, June 18th, 1868. J. B. OVERTON Sec'y.

RACES. RACES.

SECOND ANNUAL MEETING

OF THE

LAPORTE JOCKEY CLUB.

Commencing August 11th, 1868, and continuing five days

OFFICERS OF THE CLUB.

President, S. T. Brewster; Vice Presidents, John Conly, M. Fraiser, Mac. Haworth, John W. Thompson, John Bogys, Col. E. J. Lewis, W. T. S[illegible]; Steward, B. W. Barnes; Secretary, J. St. C. Wilson; Treasurer, A. H. Crew.

$2,000 IN PURSES.

FIRST DAY.

Running Race for purse of $500, free for all, mile heats three in five.

SECOND DAY.

Running Race for a purse of $300, free for all three-year-olds, mile heats.

THIRD DAY.

Trotting Race for a purse of $300, free for all trotters that have never beaten 2:50, mile heats three in five.

SAME DAY.- Running Race for a purse of $[illegible], free for all two year olds, single dash of one mile.

FOURTH DAY.

Running Race for a purse of $[illegible], free for all, two miles and repeat.

FIFTH DAY.

Running Race for a purse of $[illegible], free for all beaten horses during the meeting the first horse to receive $200 and the second $100, mile heats 3-in-5.

The above races to be run according to State rules; three or more to enter and two or more to start. Ten per cent entrance.

T. SMITH, Proprietor of Track.

By order of the Club,

La Porte, May 5th '68 J. St. C. Wilson, Sec'y

ABOVE:

Wrestling match advertisement in the *Quincy Union*, 1868

RIGHT:

Jockey Club advertisements in the *Quincy Union*, 1868

No mention was made of informal matches at Nelson Point by any known sources. However, the frontiersmen who mined at Nelson Creek assuredly enjoyed a friendly wrestling match around the campfire from time to time.

BILLIARDS.

Billiards was the most popular sport enjoyed by the miners of the region. Edwards noted the opening of a "billiard saloon" at Nelson Point in 1854. Like so many miners, he frequented the billiard saloons there:

> ...during the concert played 2 games billiards lost one $1...[21]
> Played a few games Billiards with T. Taylor Exp $1.50[22]

Edwards played billiards wherever he could find a table. Quincy was the other billiard center of the region where Nelsonites regularly played.

As early as 1857 Bass & Houck (in Quincy) advertised:

> Whenever you are weary of business and labor, with all their cares and vexations, and feel desirous of relaxation and exercise, and wish for pleasurable, muocent, and healthy amusement, call into the Bowling-alley and Billiard Saloon and you can be accommodated.[23]

Additionally, the Exchange Saloon offered billiards as early as 1866.[24]

By 1866 billiards had aroused sufficient interest to warrant a public explanation of the game:

> HOW THEY PLAY BILLIARDS.- As a great many people don't know how to play billiards, we make way for a description of the game from the pen of Doesticks, in order that they may remain in ignorance no longer, He says: "A game of billiards consists of punching ivory balls about on a table covered with a green cloth that looks like half and acre of meadow land with an India fence around it. The balls are punched with long wooden ramrods with wax on the little end to save the wood and leather from wearing out. You take your ramrod and some chalk on the table end, and then you squint; then you lift your leg; then fidole a little on your left hand with your ramrod; then you punch your ball. If your ball hits the other mans ball you have done a big thing, and you poke a lot of buttons that are strung on a wire. This is all there is about a game of billiards. Anybody can punch billiards- I can and may be you can.[25]

At every chance Nelsonites were quick to compare themselves with Quincy. A billiard saloon set the stage in a Nelson Point theatre production which addressed the subject. Over a game of billiards the players commented, "The brandy is really delicious, the billiard table superb, and the saloon a perfect palace. There is nothing in Quincy to compare with them."[26]

QUOITS.

A derivative of horse shoes, quoits was enjoyed on a regular basis by the miners of the region. Two stakes called "hobs" were driven some distance apart and as in horse shoes they were used as targets for the quoits. The quoits were iron discus-shaped rings with a notch cut in for the thumb. The rules were similar to contemporary horse shoes.

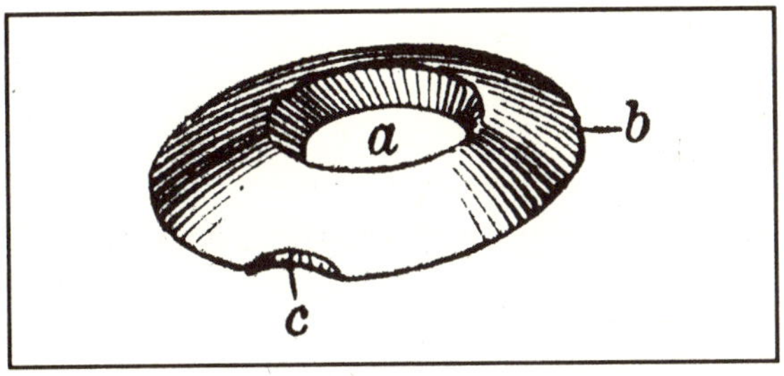

Quoit: *a*, central opening; *b*, marginal edge, which, when the quoit is skillfully pitched, cuts into the earth; *c*, thumbnotch, by which the thrower is enabled to give the quoit a spinning motion on an axis at right angles with the marginal edge.

Edwards cited many instances of quoit playing, indicating that Nelsonites often played quoits to pass the time while laid over on a trip to the American Valley at the New England or Illinois Ranches:

> Tuesday June 6th [1854]
>
> Lounged about the house [Illinois Ranch] hoed a little in the garden pitched quoits...
>
> Thursday June 8th
>
> While away the time by piching quoits, reading. etc...[27]

And in 1856 at the New England Ranch:

> Spent the day about the house joining the various games with wich the boys amused themselves (pitching Quoits, Dollars, & hunting Squirrels.)[28]

CHESS.

In 1854, Haun noted that the difference between Quincy and Nelson Point was that the men entertained themselves in Quincy with chess and in Nelson Point with fistfights.[29] Though Haun's observation was often correct, it should be noted that chess was occasionally enjoyed by Nelsonites as well. His son John, for example, enjoyed the game, "...Phillips & his partner came up & staid the rest of the day & played chess cards &c &c ..."[30] And Edwards indicated that chess was played regularly at the Sherwin House at Independence Bar.[31]

CRIBBAGE-CARDS.

Edwards recorded the popularity of cribbage in the region. During a visit to the New England Ranch in the American Valley he noted, "A.M. Snowed; Remained in doors & played cribbage to pass away the time"[32] And he played cribbage at Nelson Creek as well:

> ...the ladies all & several gents came from the Point we danced & sung & played cards till midnight
> Evening played a few games at Cribbage...[33]

COCKFIGHTING.

The *Quincy Union* labeled but one item as "SPORTS" from 1850 to 1868—cockfights. In June of 1868, The *Union* reported:

> SPORTS- We hear it reported that several trials of courage, strength, endurance and science between the best "birds" in the county will come off in Quincy on the 4th of July. A number of matches have already been made and the money put up. It is expected that arrangements will be effected to fight a main- Indian Valley against American Valley game chickens- at the same time.[34]

The origin of cockfighting in Plumas County is unclear. However, cockfighting did become a common form of entertainment throughout the camps. Haun and Edwards observed them and later Haun himself raised competitive birds.[35] By 1869, the popularity of the competition warranted the publication of fight rules.[36] These rules appear as Appendix XII.

OTHER ENTERTAINMENT.

Nelsonites attended many forms of entertainment that are difficult to categorize. Therefore, I offer the following miscellaneous examples.

In 1855 a bull and bear fight was part of the 4th of July celebration in the American Valley.[37] And Edwards noted that Nelsonites attended Risley's Circus at Elizabethtown which was considered, "rather a poor affair..."

Entertainers of all sorts toured the gold camps. Nelson Point appeared to have attracted the most "interesting" entertainment available. In 1855 a touring Miss St. Clair performed an, "exhibition of Necromancy [magic, sorcery] displaying the magic handerkerchief the inexhaustible bottle & several other magic feats..."[38] The peculiarity of the entertainment occasionally led to an unintentional show in itself:

> P.M. Prof Pinkham lectured on Phrenology[39] at Nelson House in his remarks he used language considered improper in the presence of Ladies who left the room after

which the lights were extinguished without distinction & flour thrown in his face and some of his paintings were taken to the billiard sallon and pawned for grog Dr V [Vaughan] got up a burlesque painting and himself with a ? another with a tin pan two torch bearers and another bearing said painting paraded to the different houses while Prof Pinkham undertook to escape at the back door but the boys were wide awake brot him back lodged him in a bed room and thus ended the Phrenological lecture of Prof Pinkham[40]

But before Nelsonville is renamed, "kooksville," let me end this section by noting the existence of a more civilized form of entertainment—the debate club:

> Attended a debate club at the Point the subject discussed was "Which has greatest cause of complaint against the Whites the Indians or Negro" which resulted by the president J.H.W. Nabb deciding in favor of the Indians[41]

Edwards joined the club and attended a number of meetings at the Point:

> John Cooleyham & I were appointed as principals to debate the subject of Whether the acquisition of California has been a benefit or injury to the United States
>
> Cooleyham chose the side of "Injury" & I took that of "Benefit" He chose B.B. Stevens, E.H. Sterling & W.F. Whitney as assistants & I chose J.H. McNabb Mr. Bowen & Joseph Howland[42]

DANCING.

Early in the rush, dancing was impractical due to a shortage of female dance partners. Haun made note of this on a number of occasions between 1854 and 1859. For example:

> Aug 54
> Fri 18 ...There was
> a stag dance came off of seven
> men the other hapend to be a
> gambling woman kepd up
> for some time, old married
> men engagd in the sport but
> there wives & children in the states,
> Dr. Vaughn is one of them sort.[43]

By the mid 1850's women inhabited the region in sufficient numbers to hold regular dances at Nelson Creek. Edwards noted:

> Evening attended a party at the house of Osborn Madden 15 ladies & about 100 gents were present the music was poor but the dancing good...[44]

Like his father, young John Haun regularly attended the Nelson Creek dances:

> ...going to the party at Independence. we all started in the evening & walked upon the floom we had quite a nice party at Lewis. Fox & I staid all night with Pickett I danced until I was tired of dancing)[45]

Dancing became so poular that dancing school was held at Nelson Point on July 1, 1854, presumably in preparation for the upcoming 4th of July celebration:

> Dancing school commenced (at Nelson House) taught by Millington & Co.[46]

Edwards recorded that 4th of July celebration and dance:

> P.M. A large company of persons including several ladies headed by a band playing the Russian March several of the gents being dressed in immitation of the Russian uniform and other tandry trappings with badges of blue & red ribbons bearing the words Russian Army marched up Sherwins floom to Independence Bar where a temporary building had been erected for the occasion Mr Buckbee delivered an oration very appropriate followed by a Mr Farleigh who assailed the northern abolitionists in strong language and touched on other political topics quite unappropriate to the day & had there been rotten eggs at hand he would surely have been pelted therewith We ? their adjourned to the Sherwin House and dined... afterwhich a ball was given which I joined (ticket $10.) and danced twice tome to each figure 10 minutes thus costing .50 P minute about thirty Ladies attended several of whom did not dance 106 numbers were sold the music consisted of a violin, melodean, picilo & cornette...[47]

That miners paid to dance with the few women who attended the Nelson Creek dances was not unusual. Three years later Edwards noted:

> Evening. All hands went to the Point found two dutch girls & a boy the boy played an accordeon whilst the girls danced with the miners by their paying 50¢ each for the privilege Exp $1.50[48]

Nelsonites regularly attended dances in Quincy as well:

> THE CHRISTMAS BALL- The dance at the old Court House on last Monday was well attended, and exceeded anything of the kind heretofore gotten up by MR.MASTIN. All were pleased, joyful and happy, making merry the hours as they flew...[49]

That same Christmas another ball was announced by the American Hotel, managed by Nelson Point's F.S. Fox, "Christmas Ball, at Quincy...A coitillon party will be given at the American Hotel."[50]

Conflicting evidence exists regarding the social acceptability of dancing at Nelson Point. Edwards, Hambly, and the Hauns all wrote with enthusiasm for dancing. And Nomen Nescio in his November 26, 1865, correspondence expounded on how Nelsonites passed the long winter evenings, "debating, singing and dancing schools are all in the programme."[51] However, just two months later Nescio wrote that no Christmas ball was held because of the orthodox majority. "Dancing is scarcely thought of, and condemned 'in toto' by all but four of five of the ladies."[52] Apparently, dancing was popular throughout the region with all but a few fundamentalists.

THEATRE.

Theatre productions were staged by either local performers or by professionals who toured the camps. Routinely held in Nelson Point saloons, these performances were well attended, as Edwards observed:

> Evening some 70 or 75 persons including 5 ladies attended a theatrical exhibition at the Point given by the Sager family and after which Mrs Sager played the violin & we had a dance...[53]

The success of that performance apparently prompted the Robinson family to stage a performance of its own shortly thereafter:

> Evening, nearly the whole population of this vicinity attended a theatrical performance at the Point given by the Robinson family the performance of Miss Susan was enthusiastically applauded[54]

A vivid description of Nelson Point's theatre appears in Hezekiah Bummer's January, 1860 correspondence:

> I attended our Theatre last evening. The performances were so novel, the play entirely new, and as I have the best of reasons for believing will never again be produced in the State. I have determined to give you a description. I hope I may have a full free pardon extended to me for being prosy. The title of the play was "The Grand-Mountain Steamboat- or the way to do it;" and allow me to say it was done to perfection. The audience was not large, but very select indeed, being composed entirely of the state of the beau monde. The scenery we never saw equalled in any theatre. The music I am sorry to say was an entire failure in consequence of our celebrated violinist, the man that plays the tune of the stump tailed dog so beautifully not being at his post. The pride of admission was nary "nary red," and perfectly satisfactory. The curtain rose at early candle light and exposed to our view the beautiful scenery of a store well stocked, and an actor who played the part of clerk in said store. The star actor of the evening Stranger, enters and bows very politely. He is grace and ease itself, well dressed but not foppish. He compliments the clerk and the store in an easy confidential way quite charming. Presently, all by accident, he remarks "now you don't happen to have any choice liquors on tap." The modest clerk does though- stranger prefers brandy, drinks, and remark

there is no such brandy as that in Quincy, and down goes the curtain. The curtain rises again and shows a mountain hotel as natural as life, and an hombre officiating as landlord, cook, &c..

stranger enters the hotel with a polite bow and as much grace and ease as my Lord Noodle, and orders supper. Landlord and cook is flustered, but flies to the kitchen. Bless me how the choice delicasies of the mountains accumulated on the table. Remarks no such meal as that can be had in Quincy. Cook and landlord is pleased with the compliment- Stanger regales himself with a cigar in the bar room that he says beats Quincy all hallow. The curtain now falls and when it rises again reveals a billiard saloon as natural as life. Even Hezekiah's chicken, in a cage, is dimly visible in the background. The characters here are saloon man, barkeep, and some half a dozen patron saints. There they sit toasting themselves by the side of a red hot stove with heels elevated at an angle of forty-five degrees, puffing away at a cigar of a stale pipe, dozing over a newspaper, of fast asleep. Presently one will gape, then another and another, till all that are sufficiently awake have gaped. Then one very sagely remarks that its extremely dull, and so that goes the rounds. The graceful stranger enters, bows and scrapes as no one else can do and all is life and animation where it was so dull and gloomy only a moment before. They are disconcerted a little at first, but the stranger soon makes all at ease around him. Barkeep is fully impressed with the importance of stranger, and stirs the fire desperately, and touches up the camphene lamp. Never, never did that lamp throw out such a brilliant dazzling light before. A bummer asks barkeep for tobacco,- gets a chew and then mishieviously shows barkeep a whole plug. Stranger immediately asks bummer for tobacco- takes a very delicate chew and soon desires another. Bummer in the fullness of his generous soul tells stranger to take a piece. Stranger with great dexterita cleaves the plug of tobacco at the very center and pockets half much to the discon?? of poor bummer. But it is better tobacco that can be bought in Quincy, stranger says, and that is some consolation. Saloon man treats stranger again. Stranger compliments everything this time in the most glowing eloquence. The brandy is really delicious, the billiard table superb, and the saloon a perfect palace. There is nothing in Quincy to be compared with them. Stranger banters the saloon man to play billiards, and at it they go game after game. It is now stranger and now saloon man. Barkeep fires up, and touches up the aforesaid lamp with a skillfull hand. A fall piece of humanity hour after hour wating to escort the stranger to his room at the hotel, and has become the very picture of despair hours ago, but still on, on the game goes, regardless of the waning night and fast approaching day. The game has run into the thousands- stranger has been treated time enough and finds himself sufficiently amused. He bows very low to the saloon man ways good evening, sir, and adds, in charming accents, "I will see you, sir, in the morning," and walks very gracefully out. Saloon man is stupefied for a moment. Barkeep says, "follow him." Saloon man takes three steps towards the door very quick- hesitates- stops and returns to his place beside the stove, awfully crestfallen. Stranger is wrapt in the arms of old Morpheus in the best room at the hotel, when the curtain to our great stage falls and closes the scene. I neglected to mention, but stranger modestly hinted in the early part of the evening a

> preference for an early start in the morning. The curtain rises again, and never did mortal eyes behold a better representation of morning. It reveals landlord and cook in a hurry, but with his cloth slippers on moving as quietly as a mouse, lest he shall disturb the gentle slumbers of stranger. Stranger rises partakes of his sumptuous repast and showers compliments. There is no such breakfast to be held in Quincy, oh no! Stranger says this to himself. Finally he takes a very affectionate leave, saying very sweetly, "I will settle with you, sir, when I return." Landlord and cook answers very sharply, I don't keep books, and gives stranger a frosty look that would have instantly killed a whole patch of pumpkin vines, but stranger makes his exit not in least disconcerted, and humming a pretty little tune as he goes. As he rises the mountain in the background he approaches a man sawing wood, whom he banters for a pair of snow shoes pay when he returns. The man does not wish to sell, and when stranger moves on, places his hand to his face as you have seen mischievous boys do before now, and stranger disappears.- Never was mortal man more popular than the star actor of last evening is at old Nelson. Stranger is indeed eminently qualified to travel, and has sufficient brass in his face for a lower country politician. We like him very much. He teaches a very excellent lesson. I do wonder if men will ever learn that a tasty garb and genteel bearing has in reality no more to do with cheese. We have suddenly become a very genteel people, here, if polite bows are a sure mark of gentility. But its time I closed this.
>
> Your, & c., Hezekiah Bummer[55]

Given the premise that theatre reflects the society of the players, "The Grand-Mountain Steamboat" paints an accurate picture of life in Nelson Point. And it demonstrates the need that Nelsonites felt to compare their town with Quincy.

CONCERTS.

Concerts were well received in the region, sometimes being held in conjunction with a ball or dance. Nelson Creek was a center for these events and, like the theatrical productions, both local talent and touring professionals performed. As Edwards remarked

> ...P.M. Went to the Point where a concert was given by Mr & Mrs Prisbie which was considered by many to be a humbug & themselves taken in after which a dance took place in which I participated.[56]

And the following night:

> P.M. Millington's cotillion band arrived at the Point went down to here them but was disappointed...
>
> P.M. attended a concert at the Nelson House given by Downie & Morton Exp $2.50[57]
>
> P.M. Attended a concert at Yanks Saloon Nelson's Point given by Miss Goodenow Exp $1.25[58]

In 1855, Edwards noted the performance of the "Fairy Minstrels" at Independence Bar.[59]

GAMBLING.

Undeniably, the most common form of recreation in the camps was gambling. Among the first businesses to appear in any camp was a saloon, where whiskey and gambling were offered hand-in-hand. Because of the image that Hollywood has invented, too often we picture frontiersmen sitting around a table playing poker in every saloon in the west. Although poker was played, more often faro was the game of choice.

"Bucking the tiger," or beating the odds in faro, became a common phrase and the intention of most miners at one time or another. Faro was a simple game in which one played against the house. A lay-out was used which included all the cards in one suit—usually spades. A player placed his bet on any card, ace through king. The bet rode until that card, of any suit, was dealt from a dealing shoe. The player then won or lost even-money depending on the manner in which he had placed his bet. There were a hundred or more betting variations. The most common was to bet that the card dealt from the shoe would appear as a "winner' or "loser." Because the house had only a slight advantage in the odds, the game was often played for hours without the gambler winning or losing a considerable sum. Thus, faro was an excellent means of passing the time while socializing and relaxing, following a hard day at the diggings.

Monte proved popular at Nelson Creek as early as 1850, sometimes interfering with the workaday activities. Records indicate that monte dealers dealt both at Independence Bar and at Nelson Point. Alonzo Delano described how monte disrupted mining production at Nelson Creek in 1850:

> A change had been gradually coming over many of our people, and for three or four days several industrious men had commenced drinking, and after the monte bank [table] was set up, it seemed as if the long smothered fire burst forth into a flame. Labor, with few exceptions, seem suspended, and a great many miners spent their time in riot and debauchery. Some scarecly ate their meals, some would not go to their cabins, but building large fires, would lay down, exposed to the frost; and one night, in the rain. Even after the monte dealer had cleared nearly all out who would play, the game kept up by the miners themselves in a small way, till the fragments of their purses were exhausted. There were two companies at work near me, who, when I first went there were taking out daily in each company, from one hundred to one hundred and fifty dollars. This they continued to do for more than two weeks, when it seemed as if the gold blistered their fingers, and they began a career of drinking and gambling, until it was gone...Among the

> miners who lost nine hundred dollars, another, eight hundred- their whole summer's work- and went off poor and penniless. The monte dealer, who, in his way was a gentleman, and honorable according to the notions of that class of men, won in two nights three thousand dollars! When he had collected his taxes on our bar, he went to Onion Valley, six miles distant, and lost in one night four thousand, exemplifying the fact, that a gambler may be rich to-day, and a beggar to-morrow. Gambling at that period was more prevalent in the mines than it is now; and it is but justice to say, that very many did not play at all, nor incline to dissipation; and that at this time, (1852,) a great reformation has taken place throughout the mines, although gambling is carried on to some extent.[60]

Monte originated in Spanish and Mexican card games, in which three players bet on a bottom and top layout of two cards each. A Spanish pack of forty cards was generally used with no eights, nines, or tens, although sometimes only the nines and tens were discarded, and a forty-four-card pack was used. After the shuffle and cut the dealer held the pack face down, drew the two bottom cards, and placed them on the table face up. This was called the "bottom layout." The pack was then turned up and the card exposed was the "gate." If it matched in denomination a card in either layout, the dealer paid all bets which had been made on the layout.[61]

By July of 1854, at least three establishments offered gambling at Nelson Point: the Abbot House, Fagan's Saloon, and the billiard saloon. Haun visited all three one evening:

> July 54
> Sat 29 ...After supper I & Shaw went
> down to the point to hear the
> knews. The first house we sto
> pd in was the billiard saloon.
> A woman dealing Causkinnett
> & at least 100 men 1/2 doz gambl
> lers went through with an In
> dian song & dance, then tur
> ning summer setts. We next
> came to the abbot house playing
> cards there. The Thompson house
> nothing doing but trade. Then
> Fagans saloon, a woman to
> draw custon & card playing
> afew songs & summer setts
> & making love to the lady as they call it. I tyred & went home.[62]

The following night Haun returned:

> Su 30 ...P.M. we
> went down to the point. I saw
> two woman one in each house
> dealing Causkinett & lots of
> men to bet at her bank.[63]

Other popular games in the gambling houses were the ABC game, roulette, and poker. The ABC game was played with three dice, with six letters appearing on each die. The player wagered on one letter. When that letter appeared on all three dice the bet paid triple odds.

Poker was played at Nelson Point as well. John Haun was an occasional player, "...I went to the point with him...we played poker awhile &c &c..."[64]

In its early years, Nelson Point gained notoriety as a reputed center for gambling and vice. Miners flocked to the gambling houses seeking a break from both the difficult and monotonous labor at the mines and from the solitude of the mountains. The gambling houses offered relief. But there was more to the lure of gambling. Mining itself was a gamble. To a great degree, success in mining was due to luck and, consequently, miners gambled daily on the production of their diggings. Thus, a night in the gambling houses was an extension of the gambling that occurred earlier that day at the mines.

Sport and leisure activities are expressions of the society in which they occur. The activities engaged in by Nelsonites demonstrated the values, needs, customs and beliefs of a subculture of American society. The working structures of individual sports such as skiing, bowling and billiards as well as the team sport of baseball can be compared to the evolution of the gold camp from the efforts of an individual to the work of a community. Aside from any theoretical analysis of their function, to Nelsonites these activities were simply—recreation.

NOTES.

1 Robert H. Power, *Pioneer Skiing in California* (Johnsville: 1980), p. 1.
2 Ibid., p. 2.
3 Ibid., pp. 2-5.
4 Ibid., p. 5.
5 Ibid.
6 *Plumas Standard*, "Old Nelson Took In," Jan. 28, 1860.
7 Fariss and Smith, *History* (San Francisco: 1882), pp. 204-05.
8 Edwards, *Diary* (Bancroft Library: unpublished), vol. 9, Apr. 2, 1855.
9 *Quincy Union* 5, no. 37 (Jan. 12, 1867): 3.

[10] Plumas County Museum display.
[11] Edwards *Diary*, vol. 12, Feb. 3, 1856.
[12] Ibid., vol. 6, Jn. 5, 1854.
[13] "Letter From Nelson Point," *Quincy Union* 4, no. 10: 1.
[14] *Quincy Union* 6, no. 38: 2.
[15] Ibid. 6, no. 33 (Jn. 20, 1868): 3.
[16] *11th District Fair Daily*, Sept. 25, 1890 cited in the horseracing file, Plumas County Museum Archives.
[17] Haun, *Diary* (Plumas County Museum Archives: unpublished), Nov. 17, 1855.
[18] Ibid.
[19] Hambly, *Autobiography* (Bancroft Library: unpublished), July, 1859.
[20] Ibid., September, 1859.
[21] Edwards *Diary*, vol. 6, Jn. 29, 1854.
[22] Ibid., Jn. 30, 1854.
[23] *Plumas Argus*, September, 1857.
[24] *Quincy Union*, December, 1866.
[25] Ibid. 4, no. 10 (Jan. 6, 1866): 1.
[26] "Old Nelson Took In," *Plumas Standard* 1, no. 17 (Jan. 28, 1860): 2
[27] Edwards *Diary*, vol. 6, Jn. 6, 8, 1854.
[28] Ibid., vol. 12, Feb. 3, 1856.
[29] Haun *Diary*, Dec. 16, 1854.
[30] John Haun, *John Haun Diary* (Plumas County Museum Archives: unpublished), Jan. 5, 1854.
[31] Edwards *Diary*, vol. 9, Mar. 18, 1855.
[32] Ibid., vol. 12, Feb. 4, 1856.
[33] Ibid., vol. 14, Mar. 11, 18, 1857.
[34] *Quincy Union* 6, no. 33 (Jn. 20, 1868): 3.
[35] Haun *Diary*, Sept. 24, 1854, Jly. 6, 1856; Edwards *Diary*, vol. 12, Dec. 25, 1855. Spurs used by Haun's gamebirds can be seen at the Plumas County Museum.
[36] Cockfighting file, Plumas County Museum Archives.
[37] Edwards *Diary*, vol. 10, Jly. 4, 1855.
[38] Ibid., May 25-26, 1855.
[39] Phrenology professes that mental capacity is determined by the shape of the skull.
[40] Edwards *Diary*, vol. 6, Jly. 9, 1854.
[41] Ibid., vol. 14, Mar. 15, 1857.
[42] Ibid., Mar. 22, 1857.
[43] Haun *Diary*, Aug. 18, 1854.
[44] Edwards *Diary*, vol. 9, Mar. 19, 1855.
[45] John Haun *Diary*, Feb. 29, 1856.
[46] Edwards *Diary*, vol. 6, Jly. 1, 1854.
[47] Ibid., Jly. 4, 1854.
[48] Ibid., vol. 14, May 28, 1857.
[49] *Plumas Standard* 1, no. 13 (Dec. 31, 1859): 3.

[50] Ibid., no. 10 (Dec. 10, 1859): 3.
[51] *Quincy Union* 4, no. 6 (Dec. 9, 1865): 4.
[52] Ibid., no. 10: 1.
[53] Edwards *Diary*, vol. 14, May 13, 1857.
[54] Ibid., vol. 14, 1857.
[55] "Old Nelson Took In," *Plumas Standard* 1, no. 17 (Jan. 28, 1860): 2.
[56] Edwards *Diary*, vol. 6, Jn. 29, 1854.
[57] Ibid., Jly. 3, 1854.
[58] Ibid., Oct. 14, 1854.
[59] Ibid., vol. 10, Jn. 12, 1855.
[60] Alonzo Delano, *Across the Plains and Among the Diggings* (New York: 1936), p. 153.
[61] Ramon F. Adams, *Western Words: A Dictionary of the American West* (Norman: 1981), p. 199.
[62] Haun *Diary*, Jly. 29, 1854. After an extensive search through all major archives in the state of Nevada, Causkinnett remains a mystery to me. Summer setts were somersaults.
[63] Haun *Diary*, Jly. 30, 1854.
[64] John Haun *Diary*, Oct. 12, 1856. &c was a symbol for etc.

CHAPTER XVI.

LAW AND ORDER.

...he pounded the table, he banged the law-books, he shouted, and roared, and howled, he quoted from everything and everybody, poetry, sarcasm, statistics, history, pathos, blasphemy, and wound up with a grand war-whoop for free speech, freedom of the press, free schools, the Glorias Bird of America and the principles of eternal justice!

—Mark Twain

The vast majority of miners came to California having lived in communities where law and order were well established. Consequently, they brought to the gold camps a sense of community responsibility and a respect for the law. From time to time, however, gold fever promoted greed which more than occasionally resulted in violence. Burglary, robbery, assault, battery, and murder were all documented at the mines—particularly at Nelson Point.

BURGLARY—ROBBERY.

Burglary and robbery were uncommon in the camps—even at Nelson Point. However, periodically an incident was recorded:

> Robberies, too, occasionally occurred. One poor fellow's cabin was robbed of fifteen hundred dollars while he was at work. Thus in a moment he was stripped of the result of months of hard labor. He could scarcely suspect the author of his misfortune. At a gambling house near the mouth of the creek, a man who had started for home was induced to try his luck at the monte table, when under the influence of liquor, and in the excitement of having lost his money, he attempted to sieze it again, drawing his pistol on the gambler, when the latter shot him dead! He had previously written to his family that he was about starting for home, but this one thoughtless and imprudent act cost him his life, and his family would look long and in vain for the return of the husband and father, and probably without ever learning his sad and discreditable end.[1]

ASSAULT, BATTERY, AND MURDER.

Though tranquillity generally prevailed in the camps, violence regularly erupted. Often fueled by the consumption of alcohol, claim disputes generated much of the hostility. However, the occasional senseless act of violence was responsible for marking Nelson Point as a camp more hostile than most. In July of 1854, Haun wrote:

Th 27 I & Shaw was cleaning up
bed rock all day & got gold
65$. John & his man got 5$.
After supper Shaw & I went
down to the point. There I saw
Green McHatton an old 49 min
ner from Ill, he said he knew
your fathers sister in Ill by
nane of Mrs. Clark & 4 chil
dren grown & that sche favor
the old Capt verry mutch.
my friend Shaw struck
abbot on the head twice
with a revolver hurt him
badley.[2]

Haun cited several other incidents in the camp, including one ignited by the theft of a watermelon:

Sept 54
We 6 I got gold 5 1/2$ at noon
the Johns got 18 1/2$ all day.
After dinner I went down
to the point to the election
and late in the evening two gam
blers went in to a vegetable store
& took a W. Mellon with out the
consent of the oner. They then
went back to get another, but
the owner struck one of them
on the head with a rock &
knocked him down. He got
up again & got a butcher knife
& run the proprietor out & off—
the two gamblers then turnd
& tore down the store house &
threw all that was in it to
Nelson Creek & then dard any
boddy to take it up. Ive not
seen any thing to equal that.
There was at least 100 men loogin
on.[3]

Edwards gave an account of the same incident:

> ...G.W. Ashbury & Bucklin tore down a vegetable establishment and threw it into the creek and Bucklin pursued one of the proprietors with a knife with the intention of stabbing him
>
> Thursday 7th
>
> ...a posse of the citizens arrested Bucklin & Ashbury and a meeting of the miners was called to act on their case Farley was elected Chairman & myself Vice on a division of the assembly it was ascertained that the majority were in favor of taking the law in their own hands and doing the prisoners justice immediately after a trial could be had A committee was elected to take charge of the prisoners Asbury undertook to escape was pursued by myself and others and brot back several speakers addressed the assembly afterwhich a jury was elected from the persons present and S J Milling was elected secretary and W.W. White acted as attorney for the people witnesses were examined and the alledged crime or misdemeanor substantiated (during the trial Sherrif Sharp arrived and arrested the prisoners and put them under guard) the jury after mutch deliberation presented their verdict which was as follows
>
> The prisoners to pay a fine of $300. then given in charge of the civil authority to leave Nelson Point by 8 O,clock tomorrow A.M. and not to be seen there again of if they were to receive 50 lashes each...[4]

The prisoners were escorted to Marysville for civil trial, but were discharged by Stark Esq, "on account that no one appeared to prosecute them."[5]

Frequently, Nelson Point saloons were the focus of hostility. Edwards related an incident in a Nelson Point billiard saloon in 1854:

> Milligan & Davis had a millee [melee] in which Milligan stabbed Davis & Davis struck Milligan with a billiard cue on the head the wound he inflicted had to be sewed up Davis's wound in the side was considered dangerous he had watchers through the night Milligan was given in custody of the civil authorities to await his trial on sunday next[6]
>
> ...Milligans trial was postponed to Tuesday at American Valley[7]

A little later, Haun wrote of an instance in which, "I saw a woman try to get a revolver from her paramour to shoot a gambler for braking lamp & spilling the greese all over her & cards. She cursd him & then would cry."[8] On another occasion Haun found himself involved:

> Feb 55
> Su 11 ....We cleand
> out Lawrances boxes & went t
> the cabin eat dinner & went
> down to the point. Late in the eve

ning a man by the name of Cook,
askd me to drink with
him & Vaughn (-V—said he did
not ask Mr. Haun to drink with
him. I told .V. I did not drink with
sutch a man as he was he stepd
forward one step & said I was a
damd knave, I rose up & stepd
towards him two steps, he drew a la
rge Bowey knife & said Draw
& come a head, he was Drund &
Drinking he is two cowardly to act
so when sober. I felt verry mut
ch like killing the Rogue & murd
er, but the thought of doing what
whiskey will soon do & meet a just
God. I pausd & reflected. I did not
come to this country to fight & stake
my life against sutch great odds.
He has nothing to live for but to get
drunk, this is the third time that
a deadly weapon has been pointed
at me, but not mutch sceard—[9]

In 1850, a hatchet was the deadly weapon responsible for the most infamous murder at Nelson Creek. Four accounts of this incident have been discovered:

We were startled one morning, with the report that two men had been murdered a short distance above us. On repairing to the spot, a ghastly spectacle presented itself. Two men having their heads cut open with a hatchet, lay in the creek, perfectly dead. The circumstances were these: Three men from near Vergennes, Vermont, named Ward, Lawrence, and Luther, lay in a tent on the bank of the creek, at the foot of a high, steep hill. Their bed was a flat rock, and their feet reached within a few inches of the water. As they all lay asleep, about ten o'clock at night, Ward was suddenly awaked by a noise, when looking up, he saw a man standing over him with a hatchet, in the act of striking. Instantly he sprang to his feet, and encountered another man, who made at him, but he turned and ran out at the lower end of the tent, and clambering over a pile of rocks, escaped, and continued his flight in the dark towards a cabin about forty rods distant, shouting "Murder!" Reaching the cabin, the inmates turned out as soon as soon as Ward was able to give a distinct relation of the affair; and on reaching the scene of the slaughter, they found that the assassins, after completing their work of death, had robbed their victims of about four hundred dollars each, and then had thrown their bodies in the creek and escaped. As the parties were going down, they heard the sound of somebody scrambling on the hillside, overhead, but in the gloom of the night , and the nature of the country, pursuit was impossible. Suspicion naturally enough fell on poor Ward, but an investigation being held, all circumstances were in his favor, and he was fully acquitted. Indeed, his terror, and his almost miraculous escape, scarcely allowed him to sleep for many nights. They were industrious, prudent men, and esteemed by all who knew them.[10]

Kimball Webster gave a similar account, including the first evidence of activity by Nelson Point's vigilance committee:

> ...The next camp below ours on the creek was about 80 rods distant, where three men were at work. They were all from Vergennes, Vermont, and were neighbors before they left home...
>
> On Saturday night, about twelve o'clock, one of these men came running up to our camp very much excited and out of breath, saying his two partners had been killed. His version of the matter was that, as they all lay asleep, two or three men had killed his two companions with a hatchet while they slept, and that he was awakened by the noise of the blows, to discover a man with a hatchet raised over his head, just in the act of striking him the fatal blow. He had jumped to his feet and run for his life. He had heard someone running after him, but from the sound he thought his pursuer had fallen, and after that ceased to follow him.
>
> We immediately went to the stricken camp, but could do nothing that night. The next morning we went down and found their heads cut to pieces, and also found the hatchet in the creek that was used by the murderers. The men were said to have had a considerable quantity of gold, which they kept under their heads. The gold was not to be found.
>
> We rolled the murdered men in their blankets as well as we could and buried them on the side of the mountain.
>
> It would perhaps be natural for some to suspect the survivor of the three men as being the guilty murderer of the other two, and that seemed to be the case to some extent, especially by those who knew but little of the circumstances, while all those that were present when the murdered men were found and buried, were of the opinion that he was inocent of the crime.
>
> On the next Sunday, one week after the two dead men were found and buried, quite a company of men came into our camp, all being strangers. Soon after they to inquire in relation to the circumstances accompanying the murders, and all such circumstances and conditions were minutely gone into so far as was known.
>
> After learning all that seemed to be possible to be known about the sad matter, they quietly departed. We did not know their mission at the time. Soon after we learned that they came from a small mining town [Nelson Point], a few miles distant. They heard of the murders, and of the circumstances of the three men being together, two of whom were killed.
>
> They had talked the matter over and discussed the circumstances attending them, so

far as they could learn them correctly at such a distance, and a public meeting. They had arrived at the conclusion that the survivor of the three men was undoubtedly the murderer. The men that came to our camp had been selected at the meeting for the purpose of investigating the case, and to try the supposed murderer before Judge Lynch, and if found guilty, execute him.

When they came over Sunday morning they supposed it would all be completed and that they would return to their homes before night. But after an investigation had been made by them, they came to the conclusion that the companion who had made his escape was innocent.[11]

The story appeared in the *Marysville Herald* within a week:

Murder and Robbery.

EDITOR HERALD:—A most fiendish murder was committed on Saturday night last, at Nelson Creek. The circumstances as related to me by an eye witness, were mainly, as follows:

Three men at work on Nelson Creek, seven miles from its mouth and half a mile above the forks, had retired to rest, and between the hours of nine and ten one of the three was awakened by the blow of an axe, or hatchet upon his companions; he sprang and run, closely pursued by one of them, who slipped upon some smooth stones and fell; by this means he escaped and gave the alarm to the inmates of the nearest camp some forty rods distant; a general alarm was sounded, and all in the vicinity turned out with torches and lanterns; the bodies of the murdered men were found in the Creek, their heads horribly mangled, and their pockets rifled of their contents, supposed to be about four hundred dollars. Search was immediately made for the assassins but owing to the darkness of the night were unable to gain any trace of them. With the earliest dawn of day it was renewed. Small parties started out upon every trail, in every direction, but up to the time of my leaving the vicinity had gained no clue. I had the most positive assurances that no efforts would be spared to ferret out the murderers, and bring them to punishment, and from the determined character of their pursuers I have great confidence that they will be caught and punished as they deserve.

The names of the victims were James Lawrence and Wesley Luther, both of Vermont. The person who escaped says he saw but two, and is sure they were white men.

Being near the scene of the murder and gaining my information from dozens of eye witnesses of all that transpired after the act was committed, you can rely upon all the facts stated.

L.W. RANSOM.

Marysville, Sept. 11th 1850.[12]

William Parker provides the final account:

We hear occasionally of murders and other depredations. Two men were murdered on Nelson's Creek a few days since, and robbed. Three of them were sleeping side by side, when two murderers approached with tomahawks and cleft the skulls of the two outside sleepers; the middle one sprang to his feet and escaped. He soon alarmed about two hundred men who repaired to the spot and found the bodies robbed and thrown in the river.[13]

And there were other documented murders at Nelson Creek. John Haun wrote of one in September of 1854:

> ...Lloyd & I went to the point after supper there was a man killed the night before from being stabbed...[14]

The following month, Nelson Point boasted yet another homicide, as Edwards remembered:

> ...P.M. an altercation took place at Nelson Point between John G Baxter & Patrick Taft which resulted in Baxter stabbing Taft from the effect of which he died in a few minutes. Baxter was arrested by the vigilance committee[15]

The next day Fegan was summoned:

> J.J. Fegan was summoned before the vigilance committee and proved to be the instigator of the aforesaid affray and also an arrant scoundrel and a committee of three persons were appointed to take charge of his effects and give him 10 days to settle his affairs and leave the Creek no more to return...[16]

However, Fegan did not leave. Instead, he later made charges against the vigilance committee which ultimately forced them to disband.

Claim disputes were routine at the mines. For some men, life became a seemingly endless series of disputes over boundaries, water rights, partnerships, and the like. Such was the case with Haun. He was a typical miner: he settled some disputes in court, but often decided the matter on the spot:

> Oct 54
> We .... I had som
> 25 wordes with a young man this
> morning about a claim. I told
> him that he should not work
> it. I told him I could whip him
> he said I had better try it I starte
> d towards him, he drew a revol
> ver & said I must not come two
> close to him as I had a hatchett.[17]

Many claim disputes, however, did not end in a stand-off, as Haun noted:

Sept 55
Sun 30 ....After breckfast we went to the re
servoy & examined the drift that
the intruders is making & then went dow
n to the diggins set & mended some log
heaps & then went by the cabin that
old Lloyd stays in while there, he ca
me in. I told him we ownd 4/5 of
it he said it was a damd lye, he
drew up his rifle to cock it, as he had
it when he came in but I stuck too
close to him. He then tride to get to his
bed & get his knife but I got between
he gave back & got out of doors & I
close along side of him, he comme
nsd to hollo for the Bucks & them
other fellows, two of them came to
his assistance, he talked keene for a fight
so I handed my gun to Rains & put
up my shirt to let him see I had
no weapons. He still hung on to his
rifle. I took hold of the muzzle & sn
atchd it away from him. He then
struck me & I geatherd him & thru
him in a bunch of brush, his
thumb was soon between my ivory
& my fingers in his eyes, he sung
out for help but Rains & John would
not let me be taken off him till
he sung out agian & gave then Rai
ns took me off him went in the
cabin & got his butcher knife & got to
the door by this time I had my rifle
cockd & invited him out, in a pigs a,
he did not come, so we went home
& took dinner after a while then
went down to the point. There I
saw the old cock again, he commensd to a
buse me again so whack I took
him over the head & blead him good
& had him lying on his back across
the counter, when I was pulld off
him. So we closd except a few thrusts

with the unruly member, him wi
th gaugd eyes & a bad chewd thumb
& a bad cut on the head & I barly
scratchd............
I hope God will pardon me for to days conduct.[18]

A claim dispute later developed between Haun and Terwilleger:

Feb 26
Tu 12 ...when we three got to old Terwille
gars diggins, he T stopd us by asking
what it was that we wanted to have
a law suit about. & said we had ac
ted Damd rascally in turning our wat
er out of our own ditch. This he repeated
a second time. I stepd up to him & said
I was one of the compy & he must not
say so. He repeated it again. I told him
he was a liar. He struck me with a
shovel. His son & Larison tride to do
the same, Dave fought the old man
I ran Larison with a dirk knife
and was trying to stab the old man
& Fox caught me. Dave was on him
a whipd him badly. The fuss over we
started again for home.[19]

Such disputes were not limited to protecting one's interest from an outsider. Sometimes disagreements arose between partners. When two of Haun's partners, Lloyd and Shaw, requested that they settle finances in order that they might move on, all hell broke loose:

March 55
Tu 6 ...Shaw told me early
that Lloyd wanted to have
a settlement. I was making
a stool when him & Shaw
came over. He said he wan
ted a devide of the money, we
have taken out while he was
on a spree. I told him I wo
uld not be bothered with him
longe, he abusd & threatend
so we struck at each other.

> Shaw caught me & Lloyd
> pickd up the drawing knife
> that was on the table at his
> sid & made several attempts
> to strike me with it. I got
> hold of the hatchett near the
> door he then laid down the
> drawing knife. I then went
> up to him said if he would
> go out & give me a fair fight
> I would treat him as a gent
> leman. He then went off...[20]

JUSTICE.

That was not the end of the dispute. A Nelson Point court tried the case. The week following the fracas was filled with antagonism, legal papers were served and the miners' judicial system was put into play:

Mar 55
Tu 13 ...he read his documents to
me a suit for 200$ & dam
ages for 300$ to be tride on thu
rsday 15 th...

Th 15 ....After
noon all five including Shults
went down to the point to try
another Law suit this time with Lloyd
before the jury was had some
of them expressd a wish the case
would be put off unntill tomor
row, the court concurd in that
wish so there was no objection
made.........There
is a ball to be at the point to kt
that is the reason the case was put off.
Fr 16 ...This morning we all went
down to the point, but a sho
rt time befor court was calld.
A jury was had of 12 men.
Some half dozen was objected
to & others not so good filld their

places. Shaw examind at gre
at length by the plantiffs attr
ney. Adjurnd for dinner 1$ each.
Witness stated that Lloyd offerd
me 200$ to let him continue
as he had een working but
the Jury brot in a verdict aga
inst me for 300$ & restore him
again in the company. So I
paid one of my lawyers 25$...
Lloyd was heard to say that
Hawkins said I stole gold out
of the pan. I then swore
out a peace warrant against
Lloyd for trying to cut me with
a drawing knife. Sutch is
my troubles to day & but few
to say I am right.

Sat 17 ...The constable met
them & servd a notice on Shaw & came
on & took John with him and
askd me if I was not going dowm
I said not. Lawrence came up 1
for me & said the squire was wa
iting for me. I then sent down
Lloyd was in custody. I had swo
re out a peace warrant against him.
I Shaw & John all testafide that
he tride to strike me with a Dra
wing knife & that he had killed a
man in Ky in 43, besides the
threats & insults he has made.
The squire bound him over to keep
the peace 6 months & give security
for 250$.

Mon 19 ...We all five were at work again.
I, Shaw & John got gold 13$. Lawrence
& Shults about the same but not weighd.
Lloyd is trying to get a man to work
with us but has not succeeded as yet.
We all went down to the point to
hear the case argued for a new trial
but the Judge has gone to a ball.

We 21 ...Quit early after su
pper we all went down to the
point to try for a new trial
but the squire would not grant
it. So my counsil gave notice
we would take an appeal.[21]

Such was justice in "circuit court." Circuit courts were employed regularly throughout the county as a means of settling grievances. By the mid 1850's several elected justices traveled the region, the most prominent being Squire Thomas D. Bonner:

> Squire Bonner put his "justice shop" on wheels, metaphorically speaking, and traveled from camp to camp in search of controversies upon which to adjudicate and collect the necessary fees, Upon all such journeys he was accompanied and fortified by a book of some kind, which served not only as a badge of authority, but a book of reference and a convenient substitute for the holy scriptures upon which to swear witnesses. Thus equipped, he made his appearance one day at Nelson Point, and announced himself as prepared to deal out justice with a liberal hand to all who felt called upon to indulge in the commodity. There appeared before his honor one Ransmire, who sued for a writ of restitution and $500 damages against a certain party who held adverse possession of a mining claim to which he felt himself entitled. The suit was commenced, and created much dissatisfaction among the miners, who had been accustomed to adjust all difficulties, and looked upon the invasion of a migratory justice with an unfriendly eye. Let it be remarked that it was one of the inflexible rules of Bonner's court that the fees must be paid. That was what he held court for, he said, and unless the cost of court were promptly liquidated, there was no joy for the worthy justice. It was customary for him to decide against the party whom he thought was the best able to pay the costs. Good business principles would not permit him to do otherwise. Another rule of his court was to allow no witness to testify until he had exhibited his poll-tax receipt...But to return to the case at hand. As the trial progressed, his honor began to feel uneasy about the costs. The plaintiff had nothing, and so the acute justice had already determined to decide in his favor, and thus throw the costs upon the defendant, but he feared he would be unable to collect them. He therefore made an order that the defendant be required to give bonds of costs of suit, and five hundred dollars damages, so that a decision in favor of the plaintiff would be sure to gain this modern Solomon his coveted fees. The defendant was not overanxiuos to have the trial proceed, and refused to give the required sureties. Quite an indignant crowd had assembled to witness the contest, and when the mandate for the bonds was issued they became doubly indignant. A meeting was at once called, and a committee appointed to wait upon the dignified justice, and request him to adjourn his court sine die.
>
> The members of the committee, J.H. Whitlock, Dr. Vaughn, J. Bass, Dr. Lewis, and Mr. Walker, walked into the court-room, and the spokesman thus addressed the court:

> "May it please your honor, I have been instructed by the people of the state to say to you that we can find no precedent in law where the defendent in a civil suit can be compelled to give security either for costs or damages in advance of judgement."
> "I represent the people."
> "The people have nothing to do with this case. My ruling must be complied with, or the parties will be bound over in contempt of this court."
> "If this court chooses to place itself in contempt of the people," answered the spokesman, "it must take the consequences. In the name of the people, I now commend you to adjourn this court, and not convene it again."
> It was now supper-time, and the worthy justice adjourned court till ten o'clock the next morning; but before the hour arrived he was seen ascending the mountain, his legs dangling on either side of a patient pack-mule. He had a seat of justice in Onion valley, many feet higher in the air than the river, which he called his higher court, where he sat to hear appeals from his decisions in his lower courts. Here he continued the case without the presence of the defendant, and gave judgement, being unable to either enforce the judgement or collect the desired costs.[22]

Edwards noted another Bonner case:

> ...attended a suit between McGuire & Co and McDonald & Co in regards to a creek claim claimed by both parties held before Esq Bonner at the house of Root & Lewis Nelsons Point on which the Court required four days to decide, adjourned 2.30 A.M. 21st...[23]

Most cases, however, appeared to have been tried properly and included traditional legal proceedings such as appeals and changes of venue. Haun, for example, found court in session in Quincy:

> May 55
> Tu 8 Pleasant morning. Shaw Judge Heath & I started for Quin cy to court it is in session on a case that lasted all day. I & Heath took dinner at Russells 1$ each. I paid his bills for board while we staid at court...
> We 9 We then returned back to town. & then to quincy to court, my case was calld & laid over to 1 P.M. on my act of being a way. It was then calld & went in to trial, we asked a change of venue did not get it. We demurrd but over ruld by the court...

Th 10 Breckfast 1$ each. We got throu
gh with the evidence last evening.
My Lawyer spoke this A.M. &
Lloyds this P.M. & the Jury brot
in a verdict against me for 460$
damages & costs & a restitution
of his claims. Dinner 1$ each.[24]

Jury trials were an option in Nelson Point as well:

Aprl 55
Fr 6 Pleasant morning. I went out
to the diggins & worked some time
& then went down to the point
to attend a trial by Jury of the
right of Freers interest in the
ditch by Boot but he said he did
not ask for it. The sheriff levid
an execution on it in favor of
me against Freer & Vaughn. I
paid the sheriff 6$ for services per
formd against Lloyd in suit...[25]

Not all cases were settled with such expediency, however. Justice might take months and be served only after a long, drawn out series of appeals that ended in the California Supreme Court. Such was a case involving Haun:

July 54
Sat 8 We got gold 9 1/2$. The Dept sheriff
attachd my diggins & stops me
from work. I & the Johns went
down to the point & had the case
tride befor six Jurymen as
the squire would not let me
have time to get council the
Jury gave a verdict against me.
I have five days to take an ap
peal...
Su 9 Early this morning I & John
started for the American vally
up to sister Betsys. I saw Tohom
Cox the lawyer & made an ara
ngement to take an appeal to the
Count Court...

Mo. 10 I went over to onion vally
to serve a prossess on the squire
of an appeal & give bond & pay
80$ cost...

Sept 54

Mo 11 I went up to independen
ce to serve a supeny of Pat
Curley for a witness in the cas
of Freer & Vaughn...

Tu 12 John I & Mcfall went over
to Amer Vally. In due
time court was opened Waid
the Judge. Pat Hunley under
took my case in the place of
Able. Tom Cox sick. The pla
ntiff fild five different objections
but was over ruld & the case tri
de on its merits. 4 witness again
st & 2 for me. The witness heard &
case presented, decded, against me.

Su 17 ...We took
dinner at bol Russells in the A
merican vally. We then stopd at
Brdleys & saw my lawyer. He told
me the Judge had granted a new
trial with Vaughn.

Nov 54

Sat 11 ...After noon the Dept. Sheriff Neal
came & levid on my Diggins &
tools. But he left with out taking
any of the things away...

Mo 27 ...After supper Lawrence & I, we
nt down to the point, I to see if
my papers had been sent over
by the county clerk, in case
with Vaughn for an appeal.
They have not been sent.

Tu 28 I started for the court house
of plumas early. On my arrival
I found that my papers had been
given the evening before to Sam
Russell. I left the seat of court

at noon & got home some time before sun set.

Th 30 Breakfast 1$. Took dinner at Ny ranch 1$ & left my English man there...I put up at the stanfield Ranch. A distance of 35 to 40 miles to day. Supper 75 cts.—

Decem

Fr 1 Breckfast & lodging 1$. Be fore sun rise I was on the road for Marysville a distance of 19 miles whitch I made before noon...I then went to Row & Hauns office & had a consultation about my suit, in whitch they both said there would be no difficulty in gaining it for me in the Supreme Court, only I had to send back the papers & have them maid out a new on paper of a certain kind before they would be admited in that court...

Su 17 ...We sta rted for home & arrived at 1 o clock took dinner & went down to the point. The county clerk had made out the papers again on the right kind of paper my suit with Vaughn, I sent them by Everetts express to Marysville to Row & Haun...

July 55

Th 5 ...I left 30$ with Haun & Martin to pay the supreme Clerk fees...

Feb 56

Fr 8 ...That over we went down to the point. I received a remitater from the suprem court against John Lloyd, & a letter from H.P. relative to the suit...[26]

Not only were some cases not handled expediently, but often Nelsonites were simply not satisfied with the results. Such was the case with Hambly in 1854:

> ...I likewise, was unduly Lawd out of a 40 Acre back of land, that I had claimed, on what was at that time calld the commons, but Gen. Russell through the advice of Judkins, Bradley, & others to back him, contested my rights, because, they all had more land than the Law would allow them at this time, and a combination was formed to keep intruders off, & as I was a Nelson Creek Miner I had no right to crow in on farmers, it was instituted before Esq. Starks in Elizabeth Town (Betsy Gulch) I was cinched for one hundred & Thirteen dollars. costs,—besides paying my Petty Fogger. came out victorious the wrong end: no chicanery in this oh no[27]

Prior to the development of circuit court justice, miners in the region relied on "miners' court" Miners' court operated much the same as a formal court Cases were tried in the evening or on Sundays, usually in the most hospitable saloon. Proceedings began with the election or appointment of a chairman or judge and a clerk. If a criminal case were to be tried, a sheriff, jury, and two attorneys were typically selected. In most instances the court was fair and just, serving as a prime example of how well early rush camps governed themselves. The most illustrative case tried in this fashion was the "Gilson-Wilson Trajedy":

> Sometime in the fall of 1853, a man named Gilson was keeping a boarding-house on Nelson creek, at a point called Henpeck Flat, in which occupation he was ably seconded by the exertions of his wife. Women were a *rara avis* in the mountain mining camps at the time, and were the object of a great deal of attention and admiration from the crowds of the sterner sex who had so long been deprived of the pleasure of their society. Henpeck flat was no exception to this rule; but be it remarked, that the utmost respect and propriety pervaded the feelings and actions of the boarders who sat about Mrs. Gilson's bountiful table towards the lady who placed the products of her culinary skill before them. There was one black sheep in the flock, a man by the name of Wilson, whose wiles and blandishments so worked upon the feelings the lady that she left the house and committed herself to the care of the faithless boarder, and lodged no more under her husband's roof. This proceeding was not calculated to inspire peaceable feelings in the breast of the wronged husband; and a few days later his wrath was excited to fever heat by seeing through the window of a neighbor's house the faithless wife and her seducer sitting together on a sofa. Taking his pistol, the outraged husband walked up to the window and fired from the outside, inflicting a wound upon the man within, which was supposed to be mortal, but from which the man finally recovered. A dozen of Mr. Gilson's boarders were witnesses of the act, but made no effort to prevent it. A tremendous excitement soon spread up and down the creek, and a miners' meeting was called, attended by nearly a thousand men, Gilson being held in custody. It was decided to try the prisoner; and a jury of twelve men was selected, a man named Clownie

> appointed judge, and Charles Whitlock, clerk. The prosecution was conducted by William White, and the defense by Major James Whitlock, now postmaster at Quincy. The trial continued for three days; and at its conclusion Gilson was acquitted, on the ground that he had but given Wilson his just deserts. The verdict failed to inspire Wilson with confidence in his future security, and as soon as he recovered from his wound he made a hasty departure for a more salubrious clime. Mrs. Gilson was from Michigan; and friends of the family induced her to return home, her husband furnishing the requisite means. They were afterwords reconciled, and are now living together in conjugal felicity.[28]

Haun casually mentioned the incident, suggesting that it occurred in early September:

> Sept 1853
> Sun 4 ...John washd a
> shirt a piece & two towels. Afternoon
> the man that lost his wife a few
> days past shot the one who had her.
> Mo 5 He is not ded yet but is thou
> ght cannot live. The miners are
> trying the man for shooting...[29]

Most miners apparently felt that Wilson was a scoundrel and, consequently, sided with Gilson. However, the shooting occurred in Hambly's house and he placed the blame squarely on Gilson:

> I had no particular object of writing off this copy of the Nelson Creek Miners Laws but the chairman Gilson was a jealous buck & a coward he privately abused his wife at nights by pinching & other maltreatment because a Dick Wilson from the City of New York, a gentleman in accomplishments presumed to make himself agreeable in her company & all other females society, it was too much for Mr. G. & he got his green eye goggles on. his poor wife had to flee, or take the abuse from a man in physique, but not in Soul, she came to my house, as she and my wife were friendly, I appraised Wilson, in the evening of her being at my house, and requested that he would find board at some other house, he left after supper, and did not again call, until the Sunday following everything was discussed pro & con and supposed to be all right. Mrs Gilson was stopping with me, keeping my wife, as she was not as yet very strong being lately confined &c Wilson kept himself aloof as much as possible from my house, but on the 4th of Septr / 53 he came to my house to pay me for board & washing that he had not yet settled since he left; Mrs Gilson & my wife were cleaning raisins and standing very close together, while I was on the opposite side of the room making out, or drawing off bills, rocking the infant in the cradle, and passing remarks to Wilson, who was standing on the floor with his back to the window when Gilson creeped up on him, & shot him firing through the window, a row on hand the Nelson Creek clicque, in plain English the "roughs" they

were hoping they had a case, in Mrs Gilson, but they were sorely fooled I made a subscription to send her home, I collected 197. dol. with what she had, bought her ticket for N.Y. and she left Nelson Creek. & in a few weeks Gilson sold his house to Mandonald and was himself ready to follow at any moment, the man was rather to be pittied, than abused, his make up was a good deal of a low cuming & bluff, he had no honor in his constitution he would always be on the strong side, or popular one, if he could so divine it.[30]

Edwards documented other cases decided in miners court which suggest that he was frequently appointed as an arbitrator:

P.M. Seeliger loafed whilst I tried 4 suits of law at Nelson Point[31]
...Acted as an arbitrator in a difference between Bowers & Smith, Snuffin, & Phelps in regard to a division of a piece of ground held by them on Independance Bar[32]

A formal judicial system began to take shape as Plumas County organized. However, the earliest formal county cases were still tried under primitive circumstances. Just three months after the formation of the new county, Edwards reported:

Wednesday June 7th

Spencer & myself procured horses a rode to American Ranch [Quincy] the county seat of Plumas (pro, tem) the ranch keeper requested us to get horses shod County court being is session the grand jury occupied the blacksmith's shop during which time the blacksmith had to suspend business so of course our horses went barefoot...returned via county seat found several of the grand jurors and others of the court playing marbles at the court house door a loghouse the court being called to business there were three judges one occupying a whiskey keg on end as a seat one a chair the other a stool there were 9 spectators 7 of whom occupied a long bench 1 floor & myself a whiskey keg as seats...[33]

The newly formed county remained in the 9th Judicial District, which also included Shasta, Butte, and Colusa counties. During the first term of the 9th Judicial District Court, R.I. Barnett, "a miner living on Nelson creek," and a former attorney in Missouri, was sworn in as an "attorney and counselor at law of this court."[34] Barnett went on to become a prominent name in law and politics in the region.

Formal justice was often supplemented, and in some cases entirely replaced, by the will of the people. Vigilante action, although active in surrounding counties, was not prevalent in Plumas county—with two exceptions: Rich Bar (North Fork) and, not surprisingly, Nelson Point. The two camps regularly organized vigilance committees during their early years for the express purpose of curtailing the activities of their undesirables and as a result of their lack of confidence in the formal judicial system. Rich Bar's Committee of Vigilance ultimately failed by ignoring the activities of the "Moguls."[35] Nelson Point's committee, on the other hand, experienced significant success due to popular support generated by the repeated failure of the formal judicial system to satisfy the peoples sense of fairness. The vigilance committee was formed during the first few months of the rush at Nelson Creek in 1850. And by most accounts it was an effective means of controlling unlawful acts on the creek. However, as Plumas County organized and formal law enforcement was established, the sheriff became increasingly annoyed by the committee's activities. Its operations were ultimately curtailed in 1855 by county officials who found in Fegan's complaint the excuse they were waiting for to eliminate the extralegal competition with their judicial system:

> In 1853 they [vigilantes] gave notices to leave to a great many of the undesirable citizens; but as the warnings were considered salutary, and were quickly acted upon, no occassion arose for the display of violence. At one time, in 1854, a man named Peter Taffe was stabbed to death by John Baxter, under such circumstances that the vigilance committee, after due trial of the case acquitted the defendent. They went a step farther than this, and aided the man who had but just been their prisoner to elude the proper officers of the law and make good his escape. The latter act failed to please the officials, who sought a means of punishing them. The committee took action the following winter to clear the community of a few objectionable characters, and in the course of their proceedings committed some act that laid them open to the charge of robbery by J.J. Fegan, one of the men on their black-list. This was the opportunity the officials desired. Indictments for the robbery were found in February, 1855, against James Sherwin, A. Hargrave, Jonathan Meeker, Captain Hardy, James Wooden, J.S. Root, L. Cross, Ed. Sterling, M.H. Farley, M. Parker, Fred McDonald, and Sylas Aldrich, and the defendents were arrested and incarcerated in the Quincy jail. They were all quickly bailed out but Sherwin, who refused to give bonds. He lay in jail until April, when a "nolle prosequi" was entered, and all were discharged from custody. At the same time A.L. Page was also discharged, having been in durance vile on a charge of obstructing the officers when serving the warrants of arrest upon the vigilantes.[36]

Edwards became a prominent vigilance committee member. In 1854 he noted:

> P.M. a meeting of the miners convened at Picketts store to settle a dispute between O.H. Spencer & A Johnson[37]
>
> ...was notified by Maindonald that I was appointed a member of the vigilance committee[38]
>
> 7 O,clock the
>
> Committee assembled at Parker's store and after a great deal of discussion not having decided what course to pursue in regards to answering the indictment [Fegan's], disbanded. for want of unaminity defering said publication till the next meeting. adjourned sine, die, and thus ended the organization of the Vigilance Committee of

Nelson Creek which I pronounce the climax of humbugs not to be equaled even by that prince of humbugs P.T. Barnum[39]
Sherrif Sharp arrived at Nelson Point and arrested several members of the Vigilance Committee
E.W. Sterling gave bonds to appear at court to answer the indictment
J.L.C. Sherwin declared he would not be taken by him & he leftwithout arresting him
Lewis Cross said he should not resist, and went along, but declared he should return to his Blacksmith's Shop on the morrow to sharpen the tools that miners might bring for that purpose, as it would be much to their disappointment his being absent...
Sherrif Sharp was deputising a posse to go to Independance Bar to arrest Sherwin himself & some 8 others proceeded to his residence but returned without the prisoner, but before the Sherrif himself had arrived Sherwin came down in his usual business way to attend some private business affairs in defiance of him & his posse...[40]

P.M. Sherrif & posse from American Valley went to Independance Bar to arrest Sherwin he being informed at their approach retreated to the mountain side & concealed where he could witness their proceedings in searching his premises & vicinity for him when he saw their fruitless efforts etc and being on the point of leaving without him he came from his hiding place & gave himself up walked down the flume to Sterling Flat where they undertook to place him on horseback but after a scuffle finding him too much for them they gave that up he then suffered them to handcuff him & he walked on and finally mounted the horse they had prepared for him and on which they lashed him tieing his feet underneath and his hands to the horn of the saddle he refused to give bonds therefore was taken to the County Goal

A.L. Page, was also arrested on suspicion of informing Sherwin of the approach of the officers and refused to give bonds so he also went to Goal[41]

Maindonald and Ashbury were also arrested, but Judge Ward acquitted them all and ordered:

> ...that Committee go home disband & mind their own business, and a roar of laughter ensued, which was followed on the very threshold of the court house by three cheers for the highwaymen from a large crowd[42]

Haun cited vigilance committee activity as well:

Feb 1855
Sun 25 ....After
dinner I went down to the point.
I learned from Whitney that the
vigilence committy had calld a
meeting yesturday & had added 10
moore to their mumber making 25
in all & are determined to resist the
civil law in the case of Capt. Fagan

> as the Grand Jury found a true bill
> against the vig committy 3 weeks
> since...[43]

Dissatisfaction with the formal judicial system among Nelsonites lasted long after the demise of the vigilance committee. In 1864, the *Quincy Union* reported:

> OLD "PAP" AGAIN.—A few days since, this very intelligent Justice of the Peace summoned a jury, and held an inquest over the body of the man killed in the affray at the Head of Nelson. After the jury had rendered its verdict, Pap, thinking his Court was as "good as any other man's," was about to summon a petit jury, and proceed to "try" Callich on a charge of murder, and he could not be convinced of his want of authority. So the friends of the man, charged with the killing, advised him (Callich) to leave, and during the time, they would procure a warrant for his arrest, and have an examination held before some other Justice, who knew enough to know his duty as an officer. The result of the examination is noted elsewhere. It is a positive shame, that such an ignoramous should be clothed with any official authority.[44]

MINING LAW.

Federal, state, or county mining laws did not exist during the early years of the gold rush. Consequently, each gold camp developed its own regulations, which in the case of Nelson Creek, were enforced by the vigilance committee. Hambly recorded the adoption of Nelson Creek's miners' law in March of 1853:

> ...the most of the miners that went below from this section are back & there is a lively influx of immigrants from below so much that they are interfearing with our claims & making it hot; altho we had a Miners Meeting held by appointment on the 19th day of March 1853 for the purpose of adopting Laws for the better regulation of Miners on Nelson Creek. The following laws were adopted. viz. Not supposing that the copy of these laws are worth any thing, as to its merits today, but they will see, whoever may read this paper for the amount of wisdom & justice contained in a few lines as Law by Miners, in the Early settlement of our County prior to this Law, or Rules & Regulations for the miner our Laws were verbal—
>
> MINERS LAW
>
> Section 1st In taking up bank claims, it shall be the duty of the person or persons interested to put up a notice in writing with their names signed thereto, stating the no of claims so taken, said notice to be renewed every ten days for the space of thirty days, within which time the person or persons taking up said claims shall commence active operations, said person or persons shall have the right to hold two (2) claims 50 feet on the creek straight measure, running back to the mountain, shall constitute a claim, each and every claim remaining Idle for the space of ten (10) days while workable shall be forfeited unless for good reasons showed, Bank claims shall not obstruct Creek Claims by throwing there rocks, dirt or whatever else, to the determent of said Creek claims.

> Section 2nd In taking up Creek Claims it shall be the duty of the person or persons interested to put up a notice in writing to be renewed every ten (10) days with their names signed therunto, stating the No. of claims so taken, which notice shall remain good untill the Creek shall be sufficiently low for working said claim or claims, it shall then be the duty of these interested to commence opperations, otherwise there claims shall be forfeited, unless for good causes shown, as much of the Creek as a person or persons or Co shall have, turned Either by racing, or fluming from its natural channel before, interfering with other creek claims, shall constitute a claim, or not less than on hundred feet to the man.
>
> Section 3rd Each and Every person working Creek claims shall have the right to purchase his partners interest in said claims, & work as if no sale had taken place. persons holding Creek claims shall not be deprived of the priveledge of holding bank Claims as heretofore provided. Claims of prior date shall have the preferance in all cases whatever; Cos have prosecuted their claims and failed to get them worked out and said Creek during high water has worked itself into its natural channel, it shall not operate as a forfeiture for succeeding seasons.
>
> Section 4th These Laws shall extend from the mouth of Nelson Creek to the foot of Independence Bar.
>
> Section 5th Any person selling a claim, and holding said claim in security for the payment therof, shall have the said been recorded at the Recorders office herein provided for within Ten (10) days of sale, or otherwise forfeit the said lein.
>
> Section 6th There shall be a recorder appointed by the meeting whose duty it shall be to record all leins or other business pertaining to his office, and in Case of his absence he shall have the power to appoint a successor= On Motion Joseph P. Hargrave was appointed Recorder
>
> Nelson Creek
March 19th 1853
>
> Joseph H. Gilson xxxxxxxxx
Chairman
>
> Joseph P. Hargrave
Secty.
>
> These Laws to take effect from the date of this meeting
>
> Witness DWH[45]

These laws were revised in October of the same year:

> Attended miners meeting enacted laws in reference to Bank Claims[46]

Primitive circumstances on the frontier required the establishment of respect for law and order. This need was met at Nelson Point by the development of a judicial system that was most often fair and expedient. The certainty that punishment would be swift and just fostered a common respect for the law which further assisted to maintain order. For example, an exhausted Nelsonite destined for nearby Poorman's Creek wrote:

I had overrated my strength, and found I could not work. In fact, I gave out before we got to the creek and what to do we did not know, but seeing a mule that had strayed away from someone I said, "If I could catch that mule I would ride him," although I knew that if caught it would be a case for hanging...[47]

NOTES.

1 Alonzo Delano, *Across the Plains and Among the Diggings* (New York: 1936), p. 154.
2 Haun, *Diary* (Plumas County Museum Archives: unpublished), Jly. 27, 1854.
3 Ibid., Sept. 6, 1854.
4 Edwards, *Diary* (Bancroft Library: unpublished), vol. 6, Sept. 7, 1854.
5 Ibid., Sept. 10, 1854.
6 Ibid., Jly. 12, 1854.
7 Ibid., Jly. 12, 16, 1854.
8 Haun *Diary*, Aug. 4, 1854.
9 Ibid., Feb. 11, 1854.
10 Delano *Across the Plains*, pp. 153-54.
11 Kimball Webster, *The Gold Seekers of '49* (Manchester, N.H.: 1917), pp. 158-60.
12 *Marysville Herald*, Sept. 13, 1850: 2.
13 Dr. William Tell Parker, *Notes By the Way: a journal of the gold-rush days* (Huntington Library: unpublished), p. 117.
14 John Haun, *John Haun Diary* (Plumas County Museum Archives: unpublished), Sept. 5, 1854.
15 Edwards *Diary*, vol. 8, Oct. 5, 1854.
16 Ibid.
17 Haun *Diary*, Oct. 25, 1854.
18 Ibid., Sept. 30, 1855.
19 Ibid., Feb. 12, 1856.
20 Ibid., Mar. 6, 1855.
21 Haun *Diary*.
22 Fariss and Smith, *History* (San Francisco: 1882), pp. 208-09.
23 Edwards *Diary*, vol. 6, Jn. 20, 1854.
24 Haun *Diary*, May 8-10, 1855.
25 Ibid., Apr. 6, 1855.
26 Haun *Diary*.
27 Hambly, *Autobiography* (Bancroft Library: unpublished), Feb. 22, 1854.
28 Fariss and Smith *History*, pp. 210-11.
29 Haun *Diary*, Sept. 4-5, 1853.
30 Hambly *Autobiography*.
31 Edwards *Diary*, vol. 14, May 30, 1857.
32 Ibid., vol. 6, May 27, 1854.
33 Ibid., Jn. 7, 1854.
34 Fariss and Smith *History*, pp. 166-67.

[35] The "Moguls," according to Dame Shirley, were a group of drunken rowdies at Rich Bar.
[36] Fariss and Smith *History*, p. 215.
[37] Edwards *Diary*, vol. 6, May 17, 1854.
[38] Ibid., Sept. 13, 1854.
[39] Ibid., vol. 9, Mar. 4, 1855.
[40] Ibid., Mar. 25, 1855.
[41] Ibid., Mar. 29, 1855.
[42] Ibid., Apr. 3, 1855.
[43] Haun *Diary*, Feb. 25, 1855.
[44] *Quincy Union*, Spring, 1864.
[45] Hambly *Autobiography*, 1853.
[46] Edwards *Diary*, vol. 4, Oct. 30, 1853.
[47] Charles D. Ferguson, *The Experiences of a Forty-niner During a Third of a Century in the Gold Fields* (Chico: 1924), p. 102.

CHAPTER XVII.

JOURNALISM.

Unassailable certainty is the thing that gives a newspaper the firmest and most valuable reputation.

—Mark Twain

People in Nelson Point cherished news from the outside world. Word-of-mouth and letters from home supplied the news before newspapers were widely available. In 1853 Haun wrote "I scearcely see a newspaper."[1] The *Alta* reported on Nelson Creek in 1852, "Once in several weeks they get news from the world without, and that forms quite an era with them; perhaps it is sufficiently welcome to make amends for all their waiting for it."[2] A newspaper was never published in Nelson Point, so, during the early years Nelsonites relied primarily on the *Mountain Messenger* and other California publications such as the *Marysville Herald*,[3] Sacramento's *Placer Times*, the *Golden Era*,[4] and San Francisco's *Altá California*. These sources were often supplemented by an eastern paper such as the *New York Herald* or *New York Tribune*, which were typically included in a letter from home and carried to Nelson Point by travelers or expressmen.[5] This meant that the news was often weeks—sometimes even months old before it got there. Nevertheless, these newspapers were often read over and over before being passed on to fellow miners.

The earliest local newspapers available to Nelsonites were Butte County's *Record* and the *Mountain Messenger*. The *Messenger* was first published as *The Trumpet* in 1853. The *Trumpet* was initially written in Gibsonville and printed at the office of the *Marysville Herald*. Shortly thereafter, it was published semi-weekly at Gibsonville. The paper was soon renamed the *Gibsonville Trumpet & St. Louis News*, and in May of 1854 it was renamed again as the *Mountain Messenger* and changed to a weekly printing.[6] The *Messenger* was the primary source of news at Nelson Point during the period of its publication. Subscriptions were available for Nelsonites, as documented by Edwards when in May of 1854 he, "...Subscribed for the Mountain Messenger 6 mo Exp. 50."[7] Edwards often asked the *Messenger's* editor, J.K. Lovejoy, to publish the Vigilance Committee announcements. For example, in October of 1854 a copy of a certificate of exoneration proving the acquittal of Baxter in a Vigilance Committee trial was sent to the *Messenger* for publication.[8] In February of 1855, after adopting a declaration, the committee ordered it published in the *Messenger*.[9] Lovejoy was not always accommodating, however, as Edwards noted:

> P.M. Returned to the Point addressed a letter to Editors Mountain Messenger requesting them to return proceeding of the Vigilance Committee after publishing them as we wished to preserve the original documents for reference in case of immergency[10]
>
> Evening. Received a letter from J.K. Lovejoy Editor Mountain Messenger announcing his refusal to publish the proceedings of the Vigilance Committee in his paper assigning for a revision that some portions thereof were actionable & recommended their publication in circular form. which he would do & forward to the different trading posts in the county on being so authorized by the Committee[11]

In March of 1856 the *Mountain Messenger* moved to Rabbit Creek, where it was published until it was relocated again in March of 1864 to Downieville, where it is still published.[12]

The *Old Mountaineer*—referred to by the miners as simply the *Mountaineer*—began publication in Quincy during August of 1855. John K. Lovejoy and Edward McElwain edited and published the *Mountaineer* as an independent. In 1857, the paper was sold to John C. Lewis [from Nelson Point] and James McNabb who changed its name to the *Plumas Argus*. The *Argus* was published until 1860 when it "fell into the hands of the sheriff." Later the same year, Lewis and McElwain revived the *Argus* for three months until, Lewis moved the operation to Carson City, Nevada, where he published the *Silver Age*.[13]

Two other publications originated from Quincy's *Old Mountaineer* office: the *Plumas Democrat*, and the *Fillmore Banner*. The *Democrat* and the *Banner* were political sheets, both of which ceased publication following the election of 1856. Haun referred to the Democrat:

> Oct 56
> Th 30 ...Sherwin was here last Kt &
> pd off Lovejoy for printing the
> Plumas Democrate I pd 24$
> for subscription[14]

The *Plumas Standard* offered Nelsonites a Democratic viewpoint beginning in 1859. The publication was moved to Downieville in 1863.[15]

During the Civil War, Union sympathizers at Nelson Point subscribed to the *Quincy Union*. The *Union* began publication in November of 1861 by the Union Party. After the war, W.W. Kellogg took control of the paper and published it as an independent. Kellogg moved his presses to La Porte in 1868 and renamed the paper, The *La Porte Union*. The La Porte fire of 1869 destroyed the operation.[16]

Following the management changes in the *Quincy Union*, in August of 1866, the Union Party began publishing the *Plumas National*. It became the official paper of

Plumas County, yet little coverage of Nelson Point was supplied, particularly between 1870 and 1875. The National Publishing Company was incorporated in April of 1892 and published the paper until 1918 as the *Plumas National Bulletin.*[17]

In part, news about Nelson Point was gathered by traveling correspondents. However, letters written to the newspapers from people such as Nomen Nescio supplied much of the news about Nelson Point. Nescio regularly wrote to the *Quincy Union*, supplying details of life in Nelson Point. This method of obtaining information about remote areas was common throughout the mother lode.

Like Nomen Nescio, Hezekiah Bummer relayed news from Nelson Point to the Quincy press. In his January, 1860, correspondence he best described the value placed on a newspaper in Nelson Point:

> Newspapers are above par—Let me tell you the mountain newspapers have to pass through a terrible ordeal these times. A person will catch up a newspaper, read it all over carefully and throw it down. A few moments after he will take the same paper again and commence to read just as if he supposed the paper to be constantly manufacturing more news. Reading over and over again will make even the ablest article become dull and insipid. The Standard endures this test very well. I have frequently heard it spoken of favorably and never disparingly...[18]

Newspapers operated in the gold camps much as they did elsewhere in the country. They supplied local and national news, carried the advertisements of local merchants, and notified their readers of community events. However, newspapers in the northern mother lode presented special significance. They subdued the overwhelming solitude endured by the miners.

NOTES.

1 Haun, *Diary* (Plumas County Museum Archives: unpublished), Dec. 4, 1853.
2 *Alta California*, 1852, included in *Bancroft Scraps* 51: 1-4, p. 248.
3 Edwards, *Diary* (Bancroft Library: unpublished), vol. 9, Mar. 4, 1855.
4 Ibid., vol. 10, Aug. 7, 1855.
5 Ibid., vol. 8, Nov. 19, 1854; Haun *Diary*, Oct. 23, 1853, Oct. 12, 1854, Jan. 8, 1855.
6 Fariss and Smith *History*, pp. 156, 291; Plumas County Museum journalism display.
7 Edwards, *Diary* (Bancroft Library: unpublished), vol. 6, May 22, 1854.
8 Ibid., vol. 8, Oct. 5, 1854.
9 Ibid., vol. 9, Feb. 24, 1855.
10 Ibid., Feb. 26, 1855.
11 Ibid., Mar. 4, 1855.
12 Helen W. Gould, *La Porte Scrapbook* (La Porte: 1972), pp. 63-64.

[13] Fariss and Smith, *History* (San Francisco: 1882), p. 312. Fariss and Smith maintain that the first Plumas County newspaper was the *Old Mountaineer.* However, the *Mountain Messenger* predates the *Mountaineer.*

[14] Haun *Diary*, Oct. 30, 1856.

[15] Fariss and Smith *History*, pp. 182, 312. A chronological history of Plumas County journalism displayed in the Plumas County Museum suggests that the publication was moved in 1864.

[16] Fariss and Smith *History*, p. 312; *La Porte Union*, Dec. 12, 1868: 2; Plumas County Museum journalism display.

[17] Fariss and Smith *History*, pp. 172a, 312; Plumas County Historical Society, *Plumas Memories* no. 35 (Quincy: 1970), p. 5. Appendix VIII offers a chronology of newspaper publication in the region. The *Plumas National Bulletin* was puplished as late as 1924.

[18] "Old Nelson Took In," *Plumas Standard*, Jan. 28, 1860.

CHAPTER XVIII.

POSTAL SERVICE.

Miners most often were alone in California. Their loved ones remained at home and, therefore, mail from the east was more than just news: letters from family and friends helped to ease the loneliness of their separation.

The sudden influx of population in California created unique problems for postal service to which the U.S. Post Office Department found it very difficult to adjust. The special agent sent to establish post offices at San Diego, San Pedro, Santa Barbara, Monterey, and San Francisco did not arrive until February, 1849. By November, the San Francisco and Sacramento offices had been established but the department was unable to process the tremendous volume of mail efficiently.[1]

For many years mail from the east was routed via Panama to San Francisco for distribution. In 1860, Nelsonites received word of a significant change in the mail routes:

> GOOD.—The following note from the San Francisco Postmaster to the editors of the National, shows that the eyes of the Post Office department are opened to the superiority of the Overland mail service over that furnished by the steamers. This movement certainly is cheering. It will not be long—now that the Overland route has received such favorable attention—before we shall realize the accommodations of a Daily Overland Mail:
>
> Post Office, San Francisco, Cal., Jan 17, 1850.[2]
>
> EDITORS NATIONAL—Gentleman: Will you please inform your readers that, agreeable to instructions received from the Post Office Department, I shall on and after Monday, January 23d, inst., dispatch all matter mailed at or coming to this office for the Atlantic States, by the Great Overland Route, via Los Angeles, excepting only such as may be designated by the writers to go by steamships via Panama?
>
> On and after said day it will not be necessary to endorse letters "via Overland," as heretofore, to ensure their transmission by that route. But letters to go via Panama must be so endorsed.
>
> Very respectfully,
> C.L. WELLER, Post Master.
>
> DAILY OVERLAND MAIL.—In the Assembly on the 13th inst., Mr. Welty introduced the following preamble and resolutions, which were adopted:
>
> Whereas Past experience has demonstrated the feasibility and practicability of carrying the United States mail overland from the Atlantic to the Pacific States and that

> the same can be done in shorter time and with more regularity than by ocean steamers; and whereas, the establishment of mail stations and the carrying of the mail overland, on some suitable route, will necessarily tend to secure immigration to the Pacific States; therefore, be it
>
> Resolved, By the Senate, the Assembly concurring, That our Senators be instructed and our Representatives requested to endeavor to secure the passage of a law of Congress providing for the carrying of a daily mail from some point on the Mississippi river to some point in the State of California, over such route as the contractor for carrying the same may select.[3]

Marysville became the postal distribution center for the northern camps. Express agents carried a list of patrons from camps such as Nelson Point to the Marysville postmaster who, in return, allowed them to sort out their patrons' letters. After paying the postmaster 25 cents per letter, the agents returned to Nelson Point where they usually received a dollar for each letter and 50 cents per newspaper. Conversely, Nelsonites paid the agents $1. to deliver a letter to Marysville.[4] Letters from the east were delivered to Nelson Point in approximately two months. Haun, for example, received a letter from Kentucky on April 9th that was dated February 5th.[5]

Struggling to meet the demand, the U.S. Post Office Department established the first post offices in Plumas County in 1855:

Elizabethtown .... March 2nd
Nelson's Creek .... March 30th, John C. Lewis, postmaster
Onion Valley .... August 18th
Meadow Valley .... October 3rd
Rabbit Town .... September 13th[6]

It is important to note at this time that the Nelson "Creek" Post Office was located at the original site of Nelson Point whereas another post office, Nelson "Point," was established in 1870 near the Pauly Hotel. Just a month prior to the opening of the Nelson Point Post Office, Edwards referred to the original site when he mentioned, "...the probable establishment of a mail route via Bidwells, St Louis, Gibsonville, Hopkinsville, Independence Bar, Nelson Point, Quincy, Elizabethtown & Spanish Rancho..."[7]

Mail contracts to service the new post offices were awarded to the lowest bidders. Records of these contracts are incomplete. However, Dr. S.T. Brewster contracted to carry the mail via saddle train in 1858 to Nelson Point from Rabbit Town.[8] Brewster maintained the contract after he switched to stages on the wagon road.[9]

Edwards made a number of references to the postal system. His first mention of the Nelson Creek Post Office was recorded in 1855, only a short time after the service began:

> Received at the P.O. Nelson Point, a letter from C.H. Williams...[10]

Within months he accused the Marysville Postmaster of impropriety:

> Evening, Went to Nelsons Point signed a remonstrance against the conduct of the present PostMaster of Marysville charging our express agents 25c̲ for each letter taken from that office over & above the legal postage[11]

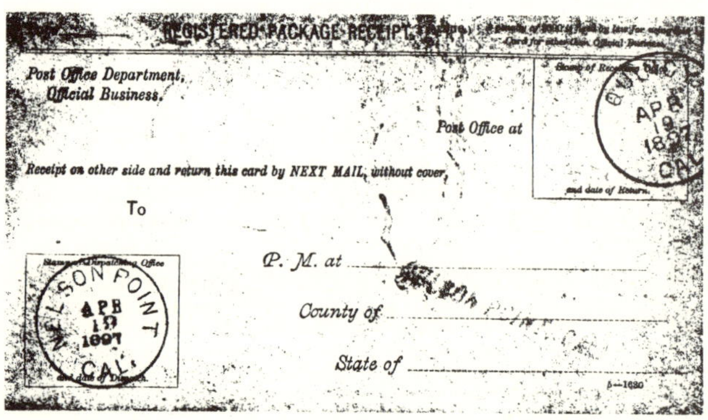

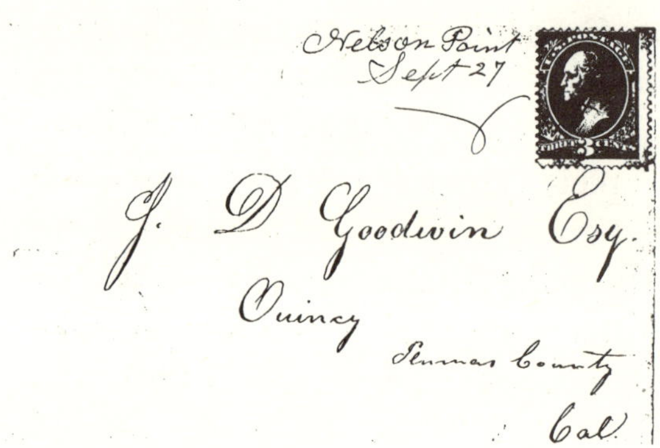

Declining population forced the closure of the Nelson Creek post office on October 10, 1859, though mail was still delivered to the camp. Deliveries were made primarily via La Porte. However, at times, mail arrived from the other direction. Again dispatched from Marysville, the route through Bidwell's Bar to Quincy was employed. In 1855, W.S. Dean delivered on this route by saddle train. He was subsequently awarded the mail contract. A twenty-five cent charge was made for letters delivered to Quincy. In 1856, Dean switched to stages. Two years later, he sold out to the California Stage Co. when the post office contracted with them to carry the mail from Oroville.[12] In order to obtain the mail from Quincy, Nelsonites formed the Mutual Express Co. Known as the "Miner's Express," the company delivered mail from Quincy once a week.[13]

The Nelson Point Post Office began operation on March 30, 1870, on the La

TOP: Registered package receipt with Nelson Point postmark, 1897.
MIDDLE AND BLEOW: Handstamp and handwritten Nelson Point postmarks, circa 1920

Porte-Quincy Road, where the bridge crossed the river. Deliveries continued by mail contract via La Porte. The Nelson Point Post Office was moved to Eclipse on July 31, 1912.[14] Mail contracts for the La Porte to Quincy run were granted to:

S.T. Brewer	1858	
Brewster & Williams	1868	
J.G. Berg	1885	
D.B. Lowery	1890	(subcontracted from Berg)
O.N. Pauly	1894	
Delahunty Brothers	1902	
Joseph Lemieux	1906	
Tom Neal	1910	
Pat Boland	1914	
Arthur T. Gould	1918	
L.B. O'Rourke	1922 [15]	

Initially, mail was delivered via La Porte to Nelson Point daily from May 1st through November. During the winters—December 1st through April—it was delivered twice weekly. By 1890, when N.O. Pauly became the Nelson Point Postmaster, mail was received tri-weekly except from December 1st through April when it was delivered semi-weekly.[16]

Winter mail delivery presented the difficulties discussed in the Transportation—Winter Travel section. They used three primary methods for snow travel. Norwegian shoe (skiing) and Knowles Snoe Shoe Express (horse snow shoe) were employed via La Porte. Whiting's Dog Express (dog sled) delivered mail to Quincy, after which one of the other methods was used to complete delivery to Nelson Point. Knowles operated only weekly, although ensuing snow-shoers would travel semi-weekly.

Roy Post was one of the last men to deliver mail from La Porte to Nelson Point. He delivered the mail to the Pauly Hotel at Nelson Point every other day during the winters of 1916 and 1917. Born in 1890, Post was a young man when he (Norwegian) snow-shoed the route. His shortcuts shaved a mile or two off the 23 mile wagon road distance to Pauly's Hotel. Roy departed about 7:00 am from La Porte and arrived at the hotel by 3:00 pm. After a hot meal from Mrs. Pauly's stove and a good night's rest at the hotel, Post waited the next morning for the mail contractor to arrive from Quincy—usually by 10:00 am. Roy then forwarded the mail to La Porte, arriving by 5:00 pm. He began his deliveries following the first heavy snow each year: one year he began on December 4th, yet the next winter he did not begin until February 1st. Along the route he encountered snow depths that varied from two or three feet at the hotel to as much as nineteen feet at Onion Valley. In order to make the best possible time, he carried two different ski dopes (waxes)—one for wet and the other for dry snow.[17]

Throughout the mother lode, the extent of postal operations indicated the prosperity of the area. When gold camps prospered, populations skyrocketed. The postal service reflected fluctuations in the populations of the region. Perhaps the single best measure of a camp's decline was the closing of its post office. The conclusion of the boom years at Nelson Creek was marked by the closing of its post office. When the Pauly Hotel burned in 1924, mail delivery to Nelson Point ceased, marking the end of another chapter of the gold rush.

NOTES.

1 Holliday, *World Rushed* (New York: 1981), p. 310.
2 The year 1850 is presumably an error and should have read 1860.
3 *Plumas Standard* 1, no. 17 (January 28, 1860): 2.
4 Fariss and Smith, *History*, p. 232. Haun, *Diary*, Jly. 9, 13 and 28, 1853, Apr. 2, 1854.
5 Ibid., April 9, 1854.
6 Walter N. Fricksted, *A Century of California Post Offices* (Butte County Library), pp. 124-5; Helen W. Gould, *La Porte Scrapbook* (La Porte: 1972), p.67.
7 Edwards, *Diary*, vol. 8, Feb. 11, 1855.
8 Gould *La Porte Scrapbook*, p. 50.
9 For additional information on Brewster see my Transportation chapter.
10 Edwards *Diary*, vol. 11, Sept. 23, 1855.
11 Ibid., vol. 10, May 26, 1855.
12 Fariss and Smith, *History*, p. 233; *Feather River Bulletin*, Aug. 12, 1937; Postal service file, Plumas County Museum Archives; Postal Route Map of 1884.
13 *Plumas National* 2, no. 24 (Feb. 15, 1868): 4.
14 Fricksted *Century of Post Offices*, p. 125; Farris and Smith, *History*, p. 288.
15 Gould *La Porte Scrapbook*, p. 50.
16 State Resources 2, no. 4 (Oct. and Nov. 1890).
17 Leroy Post, interview held at his home, La Porte, California, July 24, 1987.

CHAPTER XIX.

COMMUNICATIONS.

The first step towards modern communications in Nelson Point came when a telegraph line was completed from Sierra City to Quincy on November 16, 1874.[1] An instant link to the outside world became only a short horseback-ride away to the American Valley. As early as 1882, a telephone was connected to claims on Poorman's and Hopkins Creeks, but as late as 1892 Keddie's Map indicated that a telegraph office had yet to be established at Nelson Point.[2]

A 1912 photograph of the Pauly Hotel suggests that by that time there was a telephone at Nelson Point. An extensive search of telephone company records failed to produce any clues regarding the installation date of that telephone. However, given the accessibility of Nelson Point, compared to the remoteness of Poorman's and Hopkins it is safe to assume that the hotel telephone was installed long before 1912.

NOTES.

1 Fariss and Smith, *History* (San Francisco: 1882), p. 240.

2 Ibid., p. 321.

CHAPTER XX.

COMMERCE.

The California gold rush created a new market the likes of which the world had never seen. The sudden population growth produced a fertile ground for capitalism to flourish. And flourish it did. The demand for mining supplies: pans, picks, shovels, buckets, blankets, tents, etc., led to outrageously high profits. Food, lumber, water—the critical elements for survival and success—brought whatever price the traffic would bear. Consequently, most gold rush fortunes were earned not with a shovel, but with a ledger. The supply and demand crisis often encouraged profits of 300 to 500 percent with apples $1.50 each, pears $2.50 each, pickled peaches $16 per gallon, potatoes $1 a pound, a boiled egg 75 cents, blankets $12, and boots $35. As supplies became more plentiful and competition increased, prices became more reasonable; even then profits remained enormous.

Commerce peaked at Nelson Creek by 1855. During that year, Plumas County categorized the business licenses that were issued:

50	Liquor	License
70	Merchants	"
20	Auctioneers	"
20	Pack Pedlars	"
10	Billiard	"
10	Theatres	"
10	Concert	"
5	Bowling Alley	" [1]

A significant number of these licenses were issued to businessmen from Nelson Creek, many of whom had been operating since 1850.

Among the first to mine Nelson Creek was the Wisconsin Company. They immediately realized the profit potential in trade and opened a trading post in June, 1850, on Meeker Flat and placed W.J. Tilly in charge. They disbanded in the fall. In November, John W. Thompson built a log cabin at Nelson Point and opened a trading post. In the spring he set up a good canvas store. Between 1851 and 1853 stores and boarding houses were established at Henpeck Flat, Meeker Flat, Independence Bar, and at one or two other locations on Nelson, Hopkins, and Poorman's Creeks. It was at Nelson Point that commerce thrived, however. Nelson Point became the center of trade in the region.[2] Built on the steep banks of the creek were a hotel, a boarding house, saloons, a blacksmith's shop, a stable, and a variety of stores.

Miners traveled from as far as Rich Bar on the North Fork to secure supplies from Nelson Creek.

John W. Thompson quickly established himself as Nelson Point's most prominent businessman. Beginning in November 1850, he operated a store and bank from his log cabin. There he sold whiskey, tobacco, boots, syrup, flour, ready-made clothing, and a variety of other goods "all tumbled together in glorious confusion." Tacks were sold at $2.50 per paper, while flour, potatoes, ham, etc., went for one dollar per pound. In spring of 1851, he opened the "Thompson House" which began as a canvas store. To enable independent travel to his other operation at the Illinois Ranch in the American Valley, he operated a pack-train and rented horses from his store. Thompson sold out to Frank S. Fox in 1858. In the short time that Thompson operated, his profits were estimated to be as much as $100,000.[3]

During the boom years in Nelson Point businesses turned over frequently. Most businessmen were also miners. If his claim did not produce, the miner and his

Nelson's Creek Store, 1850

business simply moved to richer ground. Others made their fortune and returned home. The remainder of this section will describe the most noteworthy of the Nelson Point enterprises.[4] Note that some businesses located along the creek, presumably at the site of the proprietors' claims, although the center of commerce developed at Nelson Point.

The most prominent business at the original town site was the Thompson House—also referred to as the Nelson House or Nelson Point Store. It enjoyed tremendous popularity among the miners, and became the primary location for social events in the region. On Christmas of 1853 Edwards recorded that "All hands went to the Point dined at the Nelson House."[5] By 1851, Thompson had taken in two partners, John Terwilliger and J.L.C. Sherwin. In the early 50's, Frank S. Fox became both cook and operator of the place and in 1858 he bought it.[6] Edwards noted the Fox housewarming in 1855:

> Recd an invitation to attend a party to be given by F. Fox as his housewarming at the Thompson House Nelson's Point on Thursday…[7]

The Thompson House dealt in merchandise and operated as a hotel and restaurant. Fox did very well with the business for a few years, but by 1861 he owed back taxes which caused a legal summons to be printed in the *Plumas Standard*. The Thompson House was valued at $400.[8]

Sylvester Myers established himself as one of the most prominent businessmen at

Nelson Point Store.—By reference to another column, it will be seen that Frank Fox continues to supply the Nelson Point people with merchandise of every description. He has a full stock on hand, and can readily answer every demand of the mining community in which he lives.

Nelson Point store announcement in the *Plumas Standard*, 1859

Advertisements.

Notice of Dissolution.

THE Firm of MYERS & RITCHIE is this day dissolved by mutual consent,—either of the parties being authorized to settle the affairs of the late firm.
S. MYERS.
October 2nd, 1865. D. RITCHIE.

THE business will be carried on at the old Stand by S. Myers, after this date.
S. MYERS.
Nelson Point, Oct. 2, 1865. 49-tf.

Myers & Ritchie Notice of Dissolution in the *Quincy Union*, 1866

SYLVESTER MYERS,
—DEALER IN—
Groceries, Provision Liquors,
Tobacco, Cigars, Clothing,
Boots & Shoes, Hardware, &c.

NELSON POINT, PLUMAS COUNTY, CAL.
v4-n35-tf

Sylvester Myers advertisement in the *Quincy Union*, 1866

Nelson Creek. As early as 1854 he operated a saloon at Independence Bar. His saloon and restaurant were destroyed by the flood of '62.[9] He teamed with Mr. Ritchie to deal general merchandise at Nelson Point. Myers and Ritchie dissolved their partnership in October, 1865, and Myers continued to operate alone.

Myers was not the only one destroyed by the '62 flood. Others demolished included Captain Cunningham & Co.'s store and a stable and butcher shop owned by N. Pyles. The blacksmith shop tilted forty-five degrees and hung on.[10]

The Carr & Co. Store was among the last businesses to operate at Nelson Point's original location. Carr's ledger offers a fascinating look at Nelson Point commerce:

CARR & Co. STORE 1862-67

	AVAILABLE	
1862	Yeastpowder 1 doz.	4.50
	1 jar Lardoil	1.25
	Pork & Onions	
	Coaloil	.50
	Beef	51.35
1863	Nails 20 lbs.	2.50
	2 Ogres	1.50
	Sack flour	3.-
1862	Crosscut saw	10.50
	6 axhandles	2.75
	'2000' of lumber for flume	80.-
	Lumber for rockers	2.25
1862	100' water 3 days	5.-
	water 1 week & 3 nights	6.25

PATRONS (CASH ACCOUNTS)	
H. Shubert	1862
Sam Howe	1863
John Smith	'63
Dr. Carr Cash Account	'63
Flume Account	'63
Saw Mill "	'62
Dr. Carr	'62
Diederich Muicken	'63
Henry Shubert	'62
A. Long Chinaman Columbia Fl	62
Spanish Bill Foster	62
Chinamen Ling Independence	62
Henry White	63
Meyers & Richie	63
Samuel Howe	63
John Schmidt	63
Hermann	63
Charles White & Co	63
Williams & Co	63
Chinaman Squires Burr	63
David Bushman & Blair	63
James Howard & Reede	63
Chinamen China Town	63
Chinaman Aforn Buckeye	63
Chinaman Atah Rich Barr	63
Chinaman Shoon Neywood Bar	63
Gildersleeve & Guss Delerro	63
China Storekeeper Take	

A.	.63
Rich	3
Henpe	
China St	
Mitchele &	63
W. Lyons &	63
Simon Fowler	64
Johnson & Co up Creek	63
Joulan	64
Christ Beck	63
Herman Fadder	63-64
Emmons	63
Adam Wade	64
A. Riffal	64
Carr, White & Serjant	64
McGuire & Co	64
Chinaman Serv. below Mile	64
Smith Allen	64
Win. F. Reinke	64
David Crook	64
Ed McGuire & Co.	65
Hile Winders & Sherman	65
D. Bushman & Co.	64
Murcken & Reinke	'65
Capt. Geo. R. Winters	'66
John Harms	'66
David Crooks	'66
Wm. F. Reinke	'66
Kohr, Lee, Francis & Bohner	'68
Ellerman & Gohile	'66-69
F. Knock	67

'66	1 stove	14.-
	Flour & coffee	...
	Whiskey	2.25
	Vegetables	9.-
	Tea 10$.	11.25
	Rope	1.50
'67	2 Pair Gum Boots	15.-
	1 pair gloves	1.75
	1 pair Pants	1.75
	3 1/2 y Duck	1.75
	2 1/2 yards Canvass & thread	1.70
	1 overshirt	2.50
	1 blanket	8.-
	3 tobacco	3.-
	2 tobacco 2 Box Cigarrs...[11]	

The variety and availability of merchandise indicated by the Carr ledger during the 1860's did not always exist during the early years at Nelson Creek. Supplies depended on availability in the valleys and on the ability of pack trains to deliver them during winter months. And, of course, prices reflected the processes of supply and demand. In March of 1851, it was reported that "There is not a pound of any kind of provision for sale on Nelson Creek nor at Onion Valley..." And Hambly noted:

> I was fortunate in securing ample [provisions] for myself & family with vegetables &c&c [etc.etc.] Seventy five (75.$) dollars per 100$ of flour was paid at Nelson in Jan 1853 and every thing proportionally high.[12]

A boarding-house was operated in the fall of 1853 by Mr. and Mrs. Gilson at Henpeck Flat.[13]

R.I. Barnett was approved by the county to practice law on June 19, 1854, and by doing so Barnett became Nelson Point's first attorney.[14]

Plumas County business licenses were issued in 1854 to the following Nelsonites:

S. Aldridge & Co .... Nelson creek .... Merchandise, dram-shop.
John S. Root & Co .... Nelson creek .... Merchandise, grocery.
John W. Thompson & Co. .... Nelson Point .... Merchandise, tavern.
James C. Sherwin .... Nelson Point .... Merchandise, grocery.
D.W. Hambly .... Nelson creek .... Liquor.
Main, McDonald, & Co .... Independence Bar ..... Merchandise, liquor.
Myers & Co .... Independence Bar ..... Liquor.
John Mcbeth .... Meeker's Flat .... Merchandise, liquor.
William Zimberman .... Nelson Point .... Liquor, merchandise.
Bath, Powers, & Co .... Nelson Point .... Bowling-alley.
W.W. Parker .... Nelson Point .... Merchandise, liquor
Lewis & Ellis .... Independence Bar ..... Liquors.
Frank Everts, & Co .... Nelson Point .... Bank[15]

During this time commerce was at its peak in Nelson Point and Hambly gave a colorful account of his own operations:

> ...Mandonald offered me 175 dol per Mo. If I would take & run his culinary department. I told him I would for $200- but we did not accept his offer, I ran my house He got a Mrs Pearce for a little while- I put up a new grocery or Liquor stand but did not keep it open but 2 mos. then closed shop & drank my own remnants of liquor. I did not propose to supply the Dead heads of Nelson on Lick. for, or with my liquors...[16]

The outbreak of small pox at Nelson Creek in 1854 prompted the opening of a hospital. Hambly played a key role in nursing the infected back to health. As payment he apparently took a canvas covering from the hospital. This did not set right with some and Edwards reported the results:

Courtesy Yuba-Feather Historical Society

Pauly Hotel, 1912

> ...W.B.W. & myself went to Nelsonville become acquainted with Noah Greenwood late of Rome NY. attended a meeting of miners in regard to the public hospital & DW Hambly's claim thereon.[17]
> ...All hands went to Nelsonville & Point. attended a meeting of the inhabitants of Nelson Creek convened for the purpose of setting the difficulty existing betwwen D.W. Hambly & the Public which resulted in Hambly giving a receipt for $15. estimated value of the canvas covering he took from the Hospital which was considered in full of all demand & which was handed to Mr. Parsons who was elected to take charge of the Hospital till Oct 1st next...[18]

By the 1860's the rush was over at Nelson Point and, consequently, trade was in decline. When the La Porte-Quincy Wagon Road by-passed the confluence, Nelson Point commerce was doomed. The town was relocated to Lindsay Point on the wagon road, about a half mile from the bridge across the Middle Fork. There Nicholas O. Pauly opened the Pauly Hotel-Store about 1875. While he engaged in a variety of enterprises including a freight business, his wife, Catherine, operated the store and hotel. After her husband's death in 1903, Mrs. Pauly continued to operate

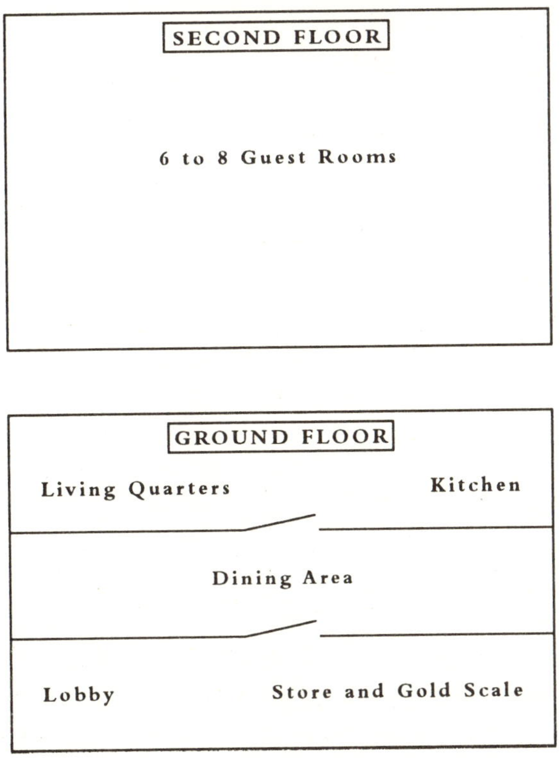

ABOVE:
Pauly Hotel and grounds, circa 1900

LEFT:
Pauly Hotel floor plan

BELOW:
Paul Hotel grounds diagram

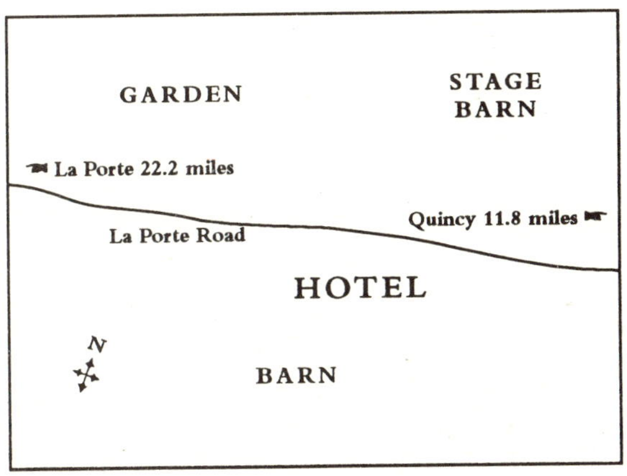

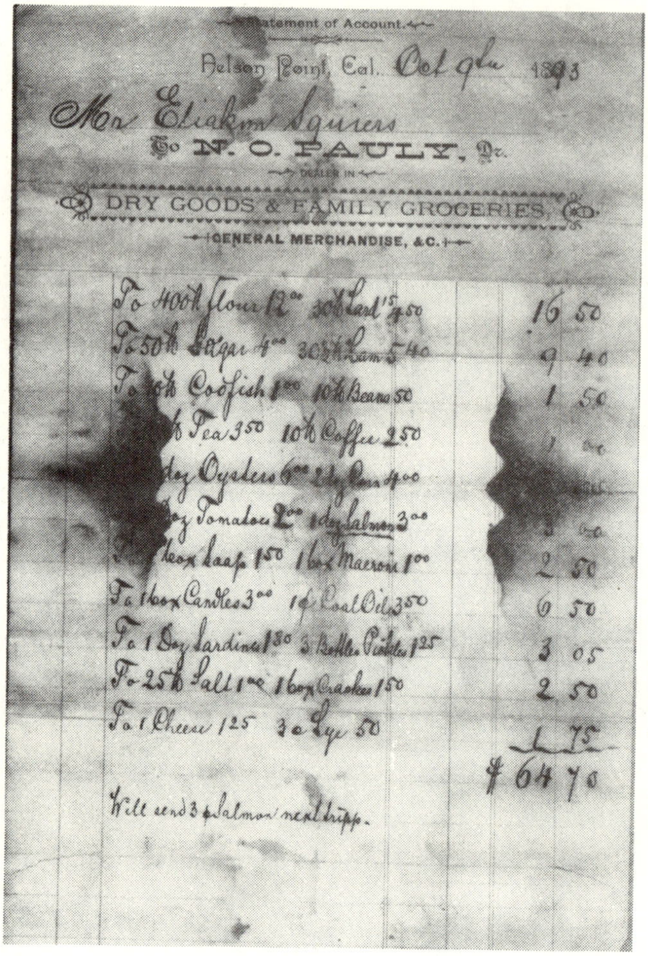
Statement of Account.

Nelson Point, Cal. Oct 9th 1893

Mr Eliakim Squires

To N. O. PAULY, Dr.

DEALER IN

DRY GOODS & FAMILY GROCERIES,

GENERAL MERCHANDISE, &C.

To 400 lb flour 12.00 30 lb Lard 4.50	16 50
To 50 lb Sugar 4.00 30 lb Ham 5.40	9 40
To [illegible] lb Codfish 1.00 10 lb Beans 50	1 50
[illegible] Tea 3.50 10 lb Coffee 2.50	[illegible]
[illegible] doz Oysters 6.00 [illegible] 4.00	[illegible]
[illegible] doz Tomatoes 2.00 1 doz Salmon 3.00	[illegible]
[illegible] box Soap 1.50 1 box Macaroni 1.00	2 50
To 1 box Candles 3.00 1 [illegible] Coal Oil 3.50	6 50
To 1 Doz Sardines 1.80 3 Bottles Pickles 1.25	3 05
To 25 lb Salt 1.00 1 box Crackers 1.50	2 50
To 1 Cheese 1.25 3 [illegible] Lye 50	1 75
	$64 70

Will send 3 [illegible] Salmon next trip.

Courtesy John D. Schmale

the hotel. All other enterprises were abandoned, for many years leaving the hotel as the only business to operate. It burned to the ground in the fire of 1924.

Commerce at Nelson Point resulted from the gold rush and, therefore, reflected the prosperity of the diggings. During the boom years of the fifties the soaring population supported a thriving commercial center servicing mines throughout the region. By the mid-fifties, Quincy's commercial enterprises began to rival those of Nelson Point, and commerce became more localized at the latter. When placer deposits in the vicinity became depleted, commerce declined until only the Pauly Hotel remained.

LEFT: **Pauly Hotel receipt, 1893**

OPPOSITE, TOP:
Pauly Hotel, 1912

OPPOSITE, BOTTOM:
Winter at the Pauly Hotel, circa 1920

Courtesy Jann Garvis

Courtesy Jann Garvis

NOTES.

1 *Plumas County Merchant Records 1854-59* (Plumas County Museum Archives: unpublished).
2 Fariss and Smith, *History* (San Francisco: 1882), p. 166-67.
3 *Plumas Standard* 3, no. 42 (Aug. 2, 1862): 1; Thompson file, Plumas County Museum Archives.
4 A more complete listing of businesses is offered in Appendix VI.
5 Edwards, *Diary* (Bancroft Library: unpublished), vol. 4, Dec. 25, 1853.
6 *Plumas Standard* 1, no. 13 (Dec. 31, 1859); Ibid., no. 10 (Dec. 10, 1859): 3; Ibid.,(Aug. 2, 1862): 1; *Plumas Argus*, September, 1857; Haun, *Diary* (Plumas County Museum Archives: unpublished), Aug. 19, 1855; Thompson file, Plumas County Museum Archives.
7 Edwards *Diary*, vol. 10, Aug. 8, 1855.
8 *Plumas Standard*, Feb. 1, 1862.
9 Ibid.
10 Ibid.
11 Carr & Co., *Carr & Co. Store Ledger 1862-67* (Plumas County Museum Archives: unpublished).
12 Hambly, *Autobiography* (Bancroft Library: unpublished), January, 1853.
13 Fariss and Smith *History*, p. 210.
14 Ibid., p. 167.
15 Ibid., pp. 203-05.
16 Hambly *Autobiography*, 1854.
17 Edwards *Diary*, vol. 6, Jan. 22, 1854.
18 Ibid., Jan. 29, 1854.

CHAPTER XXI.

FARMING.

The American Valley is one of the most beautiful in all of California...I saw turnips there which measured larger round than my waist...

—Dame Shirley

One look at the steep terrain at Nelson Point confirms that the area was not suitable for extensive farming. Only the few level parcels in the vicinity were farmed regularly. Fresh fruits and vegetables were in demand from the very beginning of the rush. Private gardens supplied some produce, but could not begin to fulfill the demand. Consequently, local ranching commenced. What was not produced locally was supplemented by products grown in the American Valley.[1] Production of wheat and barley began in 1853 in Plumas County and before long the county proved its ability to supply ample farm products.[2] The county assessor reported the following agricultural production for 1855:

Bushels of wheat	5,765
Bushels of barley	854
Bushels of oats	1,015
Bushels of corn	5
Bushels of potatoes	2,530
ton's hay	1,500
lbs. butter	2,000
horses,mules,asses	506
cattle	715
swine	1,000

By 1881, production had skyrocketed:

bushels wheat	30,000
bushels barley	7,500
bushels oats	60,000
bushels rye	6,000
bushels peas, onions, beans	2,650
bushels potatoes	4,000
ton's hay	15,000
lbs. butter	200,000
lbs. cheese	10,000
lbs. wool	30,000
lbs. honey	3,000

H. Schubert's stock on the Willow Ranch, 1885

gal. beer	15,000
fruit trees	9,150
horses and mules	2,000
cattle	10,000
sheep	6,000
swine	1,000 [3]

Nelsonites relied mostly on production from the American Valley, where three ranches were prominent: American Ranch, New England Ranch, and the Illinois Ranch. Records of Nelson Creek miners mention these locations repeatedly. For example, Edwards documented the origin of the flouring mill at the New England Ranch in November of 1854.[4]

Limited farming took place in the immediate vicinity of Nelson Point. The Willow Ranch was the only significant operation where "considerable fruit and vegetables are raised, hay being the principal product."[5] Schubert, the photographer, had a ranch near Nelson Point and ran livestock on the Willow Ranch. Later, he homesteaded the Willow Ranch and was granted title on December 2, 1890. During his homestead Schubert built a ranch house. The house was originally built on Union Creek. Schubert moved it piece by piece in the 1880's to the Willow Ranch.[6] By 1898, Schubert had sold out to his son-in-law, F. C. Mandeville. William Black

The Willow Ranch, owned by F.C. Mandeville

La Porte–Quincy Road. Willow Ranch on the right, 1905

purchased the ranch in 1912 and in 1928 he sold it to Harvey W. Egbert, Sr.

Farmers began to look for new markets as farm production in the region increased. Some saw tremendous potential, with talk in 1865 of a Quincy-La Porte Road:

> The building of the road would increase the present value of ranches, at least one half, and there is not a single one of the ranchmen but that would be making a good, yes, a big paying investment, if to build the road, he was obliged to give the one fourth value of his ranch.[7]

Completion of the road made it possible to transport fresh farm products more readily.

The demand for meat at Nelson Creek was supplied in several ways. Some Nelson Point miners raised pigs, chickens, and sheep. Edwards described some of the difficulties experienced when raising livestock at the creek:

> ...it struck a large rock which rolled down and struck the cabin but fortunately did no damage except scaring the pigs who ran for their lives"[8]
>
> ...we set a snare for a fox on the bridge another near the bridge another near the house and reset the first one on account of the great number of wild animals about & having killed 2 of our chickens the Cock & a hen that accidentally were left out of the roost yesterday evening[9]
>
> Maindonald & Co having lost some sheep on the mountains...
>
> Marshall arrived after dark bearing a sheep skin & saddle of mutton having killed one of the stray sheep & divided with the owners for his trouble[10]

The favorite meat was beef. But because of the terrain, cattle could not be successfully raised at Nelson Creek. Men like Edwards drove cattle to Nelson Creek from nearby locations. This became sufficiently profitable for Edwards to be lured away from mining for a period. In October of 1855, Edwards:

> Put a notice on Poplar Bar[11] claiming it for agricultural purposes...[12]

And on the first of November:

> Also another notice claiming Peoria Bar for agricultural purposes subject to all mining claims...[13]

Edwards delivered beef—which he often obtained from the Spring Garden Rancho—to the miners at Nelson Creek from these locations. He also drove cattle from Marysville through Nelson Point to Poplar Bar via the Illinois Ranch. He then slaughtered them and delivered the beef back to Nelson Creek. Edwards sometimes drove live cattle to Nelson Creek:

> ...I then started for Nelsons Creek [from the New England Ranch] with the Steers we purchased from Stinson...arrived at Sterling Flat at Sun Set & put up for the night after 29 miles tramp over a very rough mountain trail...[14]

Most often, however, he slaughtered them at Poplar Bar and packed in the beef. For example:

> Monday Feby 11th
>
> Left Sterling Flat went via Martin's Bar & Willow Creek carrying a pack weighing 56 lbs home engaged 3 quarters of Beef to various persons arrived home at Noon[15]
>
> I went up Nelsons Creek to Independance Bar sold 6 1/2 Qtrs Beef to the miners on the Creek...[16]
>
> Friday Feb 29th
>
> Sold out & delivered all the beef but in coming down the flume [from Sterling Flat] with a wheelbarrow load I stemmed a tree & dumped it into the flume ran ahead pitched in to get it out but lost the heart & tail which I suppose the current carried on down to the terminus[17]

The most significant farming at Nelson Point during the latter years was the limited operation of N.O. Pauly, who utilized the area surrounding his hotel for growing. In 1890, California's "State Resources" reported on the Pauly production, noting "Considerable land is devoted to raising vegetables for the hotel and hay for his stock."[18] However, this amounted to but a few acres which could not be considered as a serious agricultural operation.

In the final analysis, the vicinity of Nelson Point offered little acreage suitable for farming. Consequently, Nelson Point, like most camps in the region, relied on the American Valley to fulfill its demand for agricultural products.

NOTES.

1. California State Agricultural Society, "Report of the California State Agricultural Society," *Transactions: California State Fair and Exposition* (Sacramento: 1858-60), p. 204.
2. Fariss and Smith, *History* (San Francisco: 1882), p. 316.
3. Ibid., p. 317.
4. Edwards, *Diary* (Bancroft Library: unpublished), vol. 8, Nov. 24, 1854.
5. *State Resources* 11, no. 4 (Oct. and Nov., 1890) cited in the Nelson Point file, Plumas County Museum Archives.
6. Willow Ranch file, Plumas County Museum Archives.
7. *Quincy Union*, no. 34: 2.
8. Edwards *Diary*, vol. 11, Oct. 3, 1855.
9. Ibid., vol. 12, Dec. 28, 1855.

[10] Ibid., vol. 11, Sept. 16, 1855.

[11] Poplar Bar is located near Sloat which is approximately seven miles from Nelson Point.

[12] Edwards *Diary*, vol. 11, Oct. 17, 1855.

[13] Ibid., Nov. 1, 1855.

[14] Ibid., vol. 12, Feb. 15, 1856.

[15] Ibid., Feb. 11, 1856.

[16] Ibid., Feb. 16, 1856.

[17] Ibid., Feb. 29, 1856.

[18] *State Resources* 11, no. 34 (Oct. and Nov., 1890), cited in the Nelson Point file, Plumas County Museum Archives.

CHAPTER XXII.

MINING.

It is a pity that Adam could not have gone straight out of Eden into a quartz-mill, in order to understand the full force of his doom to "earn his bread by the sweat of his brow."
—Mark Twain

Mining began at Nelson Point as the work of individuals and developed into the work of a community. Mining was initially concentrated along Nelson Creek, where placer deposits were worked extensively between 1850 and 1851. Simple hand tools enabled individuals to operate successfully. By 1852, the surface deposits had been worked out and it became necessary to move larger quantities of gravel. Thus, more sophisticated means of mining were employed, in turn, requiring a greater variety of tools and provisions. Miners relied more and more on the "community" to provide the necessary elements of a successful mining operation.

There were many prerequisites for success at mining, but the most important remained luck. Two men operating with the same equipment in the same manner and in the same stream often produced entirely opposite results. Where one man might find nothing at all, just 20 feet away another might discover a fortune. In 1850, the *Pacific News* reported that 57,000 miners were working the Feather, Yuba, Bear, and American Rivers, most of them earning barely enough to cover expenses. Many resorted to working as laborers for day wages. Hundreds of others gave up prospecting to find work in the cities, often simply to earn a ticket home.

Early reports about Nelson Creek's riches induced hundreds of miners to rush to the area. Among the earliest published accounts appeared in a July, 1851, edition of *Alta California*:

> Nelson's Creek Diggings in 1851 must have been one of California's richest placers, They were "up on the summit of a ridge 500 feet above the ravine of Feather" Men with rockers made five hundred dollars a day. Gold lay in nuggets over the ground- good for sore eyes- and persons made good wages going around and picking them up. Here was an inducement for lazy men to make a good living in a gentlemanly way.[1]

Though rich discoveries were indeed made at Nelson Creek, most miners left empty-handed. Frank Lecouveur described his 1852 mining venture at Nelson Creek this way:

> Rothrock and his company had such poor success that he was compelled to discharge all his hired men, I being one of them. I regretted to ~~have~~ to leave these good people, to

whom I had really become attached during the time I remained with them; and there being no prospect of finding work again in the near future, and—considering the enormous prices charged for provisions—and that there is no chance for me to make a living by working alone, I packed my bundle and started back again on my way to Long Bar.[2]

Charles D. Ferguson experienced poor luck at Nelson Creek as well:

So we departed [for Long Bar], stopping on our way back to prospect on Nelson's creek for a week, but met with nothing encouraging...Our Nelson creek prospecting proved as unprovitable as our Rich Bar had, so we came to the conclusion to return to Long's Bar again.[3]

On learning of the Nelson Creek discovery, Kimball Webster and Alden J. Nutting rushed to the area. Webster wrote:

When we arrived on Nelson's creek we found a large number of people already there, provisions scarce and high, and although there were some very good mines, they did not appear to be very extensive or lasting. We made a prospecting tour farther east among the mountains for two or three days, but as we found no gold in paying quantities, we soon returned to the valley.

After we had abandoned our claim on the middle fork, I concluded to go to Nelson's Creek again...After arriving there in company with one or two others, I selected a claim on the middle fork of the Feather River, a short distance below the mouth of Nelson' Creek, where the stream could be turned by a dam. We thought there was a possibility that the river bed might be rich and concluded to construct a small wing dam sufficient to test it. We put in the dam so as to throw the water from a small portion of the river's bed. The claim proved as worthless as our first one.[4]

John Kline Kepner described the prospects for success at Nelson Creek in 1852:

Thursday 9th-

I went to town (I don't know what they call it) on Nelson Creek, and bought a shovel for $8 and one of the boys got a pick for $6. After dinner some of us went prospecting and in the evening we had about $2 of the precious stuff- quite a days work!

Some of the miners are making from 4 to 8$ per day, while there are any number of men here who can get nothing to do at any price, for the mines are all taken up and there are any number who would like to work if they could only get something to do.

What hands are employed here are getting $100 per month and board, but for one man that wants to hire a hand, there are a hundred who want the place.[5]

Hundreds, perhaps thousands, of miners came to Nelson Point. Few lasted. They arrived with visions of wealth, but often after only a few weeks experienced the harsh reality and quickly departed. Some lasted a full season, but even then only a fraction made profits sufficient to warrant establishing long-term operations.[6]

Although most miners struggled to break even at Nelson Point, some did earn substantial profits. The trick was to walk away with those profits without re-investing them in other, unsuccessful mining ventures. In this way, mining resembled gambling. Like gamblers, miners often had winning streaks, but the real winners were those who knew when to walk away from the "table."

There was no patent formula for success in a specific mining venture. But, some patterns are evident. Most men worked in small groups. First, a promising location was selected and if gold was discovered a claim was filed in order to protect the location from others. Claims sometimes covered as little as 30 feet of the stream, but more often encompassed an entire bend in the stream, ravine, or bar. Contrary to the "Miner's Ten Commandments," miners often held interest in more than one claim at the same time.[7] Edwards, for example, simultaneously filed on eight claims at Mountain Flat. Each claim was for 50 feet and carried a recording fee of $1.[8]

Partnerships formed and dissolved regularly. Consequently, in order to accommodate the transition of partners profit settlements were being made almost continually. Haun described the process:

June 54
Sun 18 John Lloyd & I went &
stakd off 7 claims for me &
John 5 we bought & 2 by pre emp
tion & pickd out one for Lloyd.
We then went up the ditch & stopd
some holes & we came home
& brekfasted. Lloyd & John went
down to the point & our knew
partners had allready gone
they
came back before noon &
divided
the gold, first paid the previous expences
of the company 39$ & then divided
the remaining 240$. making 48$
to each. I then paid 36$ for lost
time in sick nefs & then the oth
er three paid me for board 41$
& Hopkins paid me 32 1/4$ for the
use of water. I then Broke a hoe
cleaning my potatoes & went
down to the point gave 2 1/2$ for
a new one & 1/2$ for a bottle of

bitters & then came home in com
pany with Lloyd. My Lesson was
VIII chapt of Revelations.

June 1854

Mo 19 We five went to work
mad no money. Lloyd went
to the point.

Tu 20 I made a proposition to the co
mpany to take in Lloyd on the same
terms with Hopkins & it was refusd.
I then said I would not work in
this compy longer. So we did not
hing all day but disagree. Hawkins
is a mean yenkee of New York,
Shaw is a gentleman, of the west.

We 21 We mede a settlement with
Hawkins & Hopkins. Shaw bou
ght his claim in the water at
150$ & all the provisions tools
I paid him 500$ for us four
& Shaw 29$ that paid his bill.
We then went to work P.M. Shaw
went down to the point. We got
no gold worth weighing. The wind
blew pretty hard from the south P.M.

Th 22 Quite a frost last kt. Bit some of
my potatoes tops. We four all went
to work & got gold 29 1/2$. Shaw & I
work together & Lloyd & John.[9]

Entrepreneurs soon discovered that the "business" of mining offered substantial profits. By gaining interest in several profitable claims and hiring laborers to work them, skilled businessmen earned far more than most miners. Haun, for example, had interest in several claims concurrently and within several years evolved from a laborer to a managing supervisor in his own interests.

May 1855

Mo 28 Pleasant day. Shaw took the
letter down to the point & maild it
We all was mining. We got gold 42$
I shall pay off Shults he has work
d 28 days this last settlement at
75$ per 26 days, amts to 80 3/4$ &

the whole amt I paid him is
181 3/4$- to work the interest that old
Lloyd was permited to work.
Tu 29 Some clowdy & raining a verry
light sprinkle at day brake a
fine sunny day. I John & Rains
got gold 22$. Lawrance & Roister
is doing well, but Shaw, Duesler
& Judge Heath has not made any
thing, it is moore than likely that
they will not all pay at the same time.
We have an interest in all the
companys. We get 3/4 of all that
Shaw has an interest in. There
was a miner dide of bilious colic
on Sunday kt last, that lived one
mile from our cabins.[10]

Even in successful operations, daily mining profits ranged from nothing to hundreds of dollars. The following are examples from Nelson Point.[11]

The Wisconsin Co. composed of Ripley C. Kelley, Andrew Kelley, Warner Meeker, Abbey, Toy, Ivey, George Tilly, Meyers, John Warly, James Warly, S.B. Hatch, and Alex Turner, operated a claim in June, 1850, at Meeker Flat. In only three weeks they pulled out a whopping 93 lbs. of gold.[12]

In 1855, Edwards reported a single nugget taken out of the bed of Nelson Creek near Syracutus Bar that weighed 14 oz. 10 dwts. and 18 grains.[13]

Hambly arrived at Nelson Creek shortly before the onset of the first snowfall of the winter of '53. He purchased an interest in a claim:

> ...I bought (1/5) one fifth of the Buck Eye claim paid for it the first week work. My Bro. William bought likewise 1 5, on I did for him, it paid him back in a few days, we had a good piece of ground left, but it commenced to storm or rain quite severe in Nov to that degree that we had to take out our flume & wheel and lay it out on the banks of the Creek for another season. all the flumes were taken out on Nelson Creek to Meekers & cos flume, by the owners on high water. would have done it Meekers flume stood the freshet & after a few days he fixed up, hired a big crew of men and got to work; work-(2) two weeks, and then took his flume out 23 Nov 52 and with the balance of us laid up for the winter Dec Month quite cold & freezing sharp[14]

Haun recorded a daily account of his earnings which clearly depicts the range of profits that a miner experienced from day to day:

Courtesy Jann Garvis

Franklin Mine

Augst 1853

We	31	Our digging is not good this week
Sept		
Fr	2	We are not making expenses
Mo	19	All hands to diggin gold we took out today 136 1/2$...
Tu	20	Took out today 79$
We	21	" – – 58 1/4$
Th	22	" – – 176 "
June 54		
Th	8	...We got gold 41$...
Fri	9	...We got gold 68 1/2$...
Su	11	...We divided our gold, of this last weeks work, 59$ 30 cts a shear...[15]

Most miners considered placer deposits on Nelson Creek to be worked out by 1858. It then became necessary to re-work old diggings for the fine gold or to develop elaborate means of recovery in new locations. Lode [tunnel] mining became more prevalent as well.

As the white miners abandoned their old placer claims in favor of new diggings, Chinese miners often took possession of those claims. By the early 1860's the Chinese outnumbered white miners on Nelson Creek. Many estimate that by

ABOVE: Nelson Mine, 1902 BELOW: Nelson Mineshaft, 1902.
From left to right: C. Wilson, unidentified, G. Tyler, E. Metcalf

employing meticulous crevicing techniques over long and tedious hours, the Chinese recovered more gold than the white miners who preceded them. A "Letter From Nelson Point" dated June 12, 1865, gave a detailed description of the mining:

> Where river mining is the occupation of nearly three-fourths of the mining population no accounts of large yields are recorded until the latter part of the Summer or Fall, therefore the pleasure of regaling you on thousands of dollars made by Tom, Dick and Harry, must be deferred until some time after the "Fourth." The Middle Fork of Feather River for about three miles below Nelson Point is to be thoroughly cleaned out this summer. No less than five flumes besides numerous wing-dams are in course of construction between the mouth of Nelson Creek and Shasta Bar. The banks of the river present a busy appearance. The miner for the season becomes mechanic; pick and shovel are laid aside, and the ax and chisel plied in their stead until the last plank closing the dam, sends the water rushing through the flume, and then—well, he resumes the pick and shovel again and expects every day to get a small fortune to the pan, which expectation is never realized, but relieves a great deal of fatigue, caused by tenaciously holding on to a buggy which invariably upsets at the end of the road it is traveling on. Two of the claims now being flumed, one called the Fish-trap; the other owned by Maguire & Co., were worked profitably last year, and prospects obtained before the fall freshets suspended operations leave scarcely a doubt of their paying equally as well this summer. Of the others there are none but what are expected to yield every share-holder a handsome fortune, computed to suit each man's ideas of what constitutes that desirable sum.
>
> Mitchel, Giles & Co., are engaged in running a tunnel in a hill adjacent French Ravine. They are within a few feet of the shaft from which they obtained their prospect, three bits to the pan. Copeland & Co., struck good prospects, one dollar to the pan, in a cut sluiced in the hill opposite Nelson Point, unfortunately for them the water failed before the claim could be farther developed. Bonner, Hardy & Co., have been piping all winter in the old Lewis Ranch. They intend to clean up shortly. Their ground-sluice is estimated to contain about six thousand dollars.
>
> Nelson Creek is rapidly passing into the hands of Celestials. I have heard of but four or five companies of white men intending to work the Creek the coming summer...[16]

Lode mines such as the Nelson Creek Gold and Silver Mining Co. (1879) kept the industry alive and prosperous well through the 1870's, but by the turn of the century only seven mines operated in the vicinity.[17] The Nelson, Franklin, and Blue Gravel Mines were among the last to operate.

The Blue Gravel Mine developed into Nelson Points's most extensive operation. It was in full operation by 1877 and continued into the 1920's. Originally, it was the site of extensive hydraulic mining. Lode mining followed:

> BLUE GRAVEL MINE
>
> This company, whose claim is just across the Middle Fork from Pauly's Hotel is still running their tunnel from the shaft. They will raise up and commence taking gravel out

LEFT: Blue Gravel Mine, Nelson Point, mill site and buildings
MIDDLE: Blue Gravel Mine, 1920s
BELOW left: Blue Gravel Road. Aaron Lee and team

ABOVE: Mule man at the Blue Gravel Mine

> this week. The prospects of this company are bright in the extreme, and the boys deserve them, for they are all hard working men, who have a right to a fortune whether they get it or not.[18]

MINING PROCESSES.

All placer mining processes were based on the principles of density and gravity. Simply stated, gold was the most dense element in the diggings and, consequently, when mixed in the gravels, it settled to the bottom. Mining equipment was designed to expedite this process by introducing water and motion into the formula. Once the gold settled it became simply a matter of removing the gravel. Gold pans, rockers, and sluices all operated in this fashion—as did mother nature.

For thousands of years prior to the miners, gold had broken away from veins exposed by erosion in the hills. The gold then washed its way down the ravines and into the streams. The turbulence generated by flood waters, together with gravity,

William Metcalf, Sr., Superintendent of the Blue Gravel Mine

forced the gold to settle through the gravels to the bottom at bedrock. Hence most gold was found near or at bedrock.

Ferguson described one common process for extracting it:

> The gold was nuggety—some days we would get nothing—then the sinking was some six or seven feet. To strip a "paddock," as it is called (a hole), one would dig down seven or eight feet, and, perhaps get nothing; then again one might get three or four nuggets running in weight from one to four ounces, and some small or fine gold.[19]

Haun employed hydraulicing (piping) a process that utilized a large hose or "monitor" to wash the bank into sluice boxes below. This was another application of the density and gravity principles described earlier:

June 1855
St 16 I John & Tom Eaves was at work
in good time. The bank had slid
in again last kt. That kept us so
me time in getting a way the dirt
& rock. Well before noon we were
piping. I saw & pickd a piece of
gold that weighs 8 oz & 10$ just 150$
at the current rate of 17$ & 40 cts to
the oz—Shaw & Lawrance has struck
new diggins that is verry ritch, so
weve got a clain a pice there in all
4=we feel confident that we will be
able to get our piles out of those
new claims but I wont throw
a way the old ones by no means.
We cleand up some bed rock &
then the sloos boxes this P.M. &
got gold 97$ add the two to geat
her & make 235$ at 16 per oz.

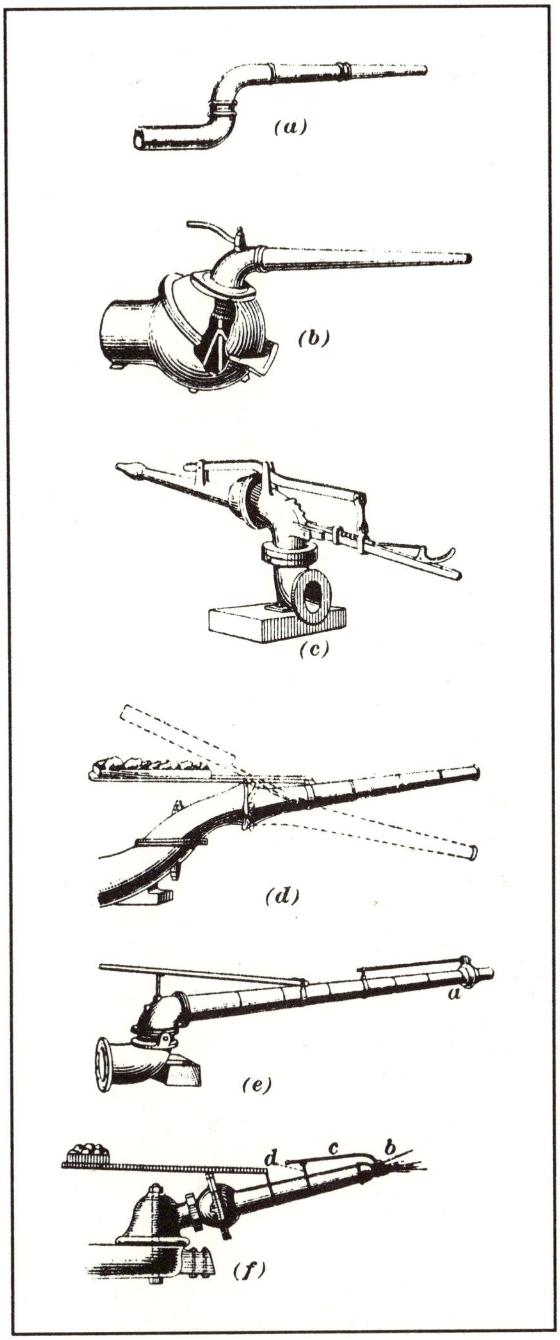

Hydraulic monitors

Augst 55
St 4 ...Well we work steady & hard all day
and only got 26$. We have so much
moving of boxes & hose & pipe.[20]

Edwards noted another hydraulic operation in 1854, "...Bowers & Co's hydraulic pipe commenced operation."[21]

In addition to operating their equipment, miners often manufactured and maintained their own devices. During the mining season at Nelson Point this became routine. One day was spent mining; the next day might require repairing tools. This routine was best explained by Haun:

Feb 1855
Sun 18 Cold this morning. I made
three Hungarian Riffles for
Lawrance & made a level. After
dinner I went down to the poi
nt maild a letter for John 25
cts to Pauline Haun & got 3 box
matches 25 cts & hired J. Shults
to fill Lloyds place
May 1854
Tu 2 The weather is verry good John
& I after getting up a little late went
out to work, we fixed for wash
ing & got gold by noon 11.$ & did
not clean over 1/2 feet square of the
bed rock. After dinner we was
ground sloosing & wet feet all even.
We 3 The weather good, we were fix
ing the boxes & mending them
in the fore noon & the after we
were washing for gold & got 19.$.
May 55
Mo 14 All hands to mining. Lawrance &
Shults to their place. John & Royster
at our old place in the ravine with
the hose & pipe. Late P.M. the hose ripd.
Shaw Rains & Heath was at work
near pikes cabin on Shaws claim.
I went down to the point to get a
man to come up & make some hose out
of No 1 duck, but did not succeed.

ABOVE: **Hydraulic mining operation**

LEFT: **Gold panners**

BELOW: **Gold pan**

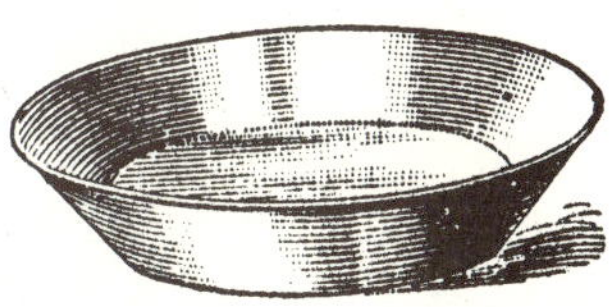

My friend Duesler came up to see after his claim p.m. He commensd making our hose. I made 4 sloos boxes out of new plank. Clowdy rai nd most of the p.m.

Tu 15 Clowdy & a light sprinkle of rain, quite cool this morning. I mended the hose, made 3 false bottoms & made a pen stock. John & Roister got gold in two days 26$.

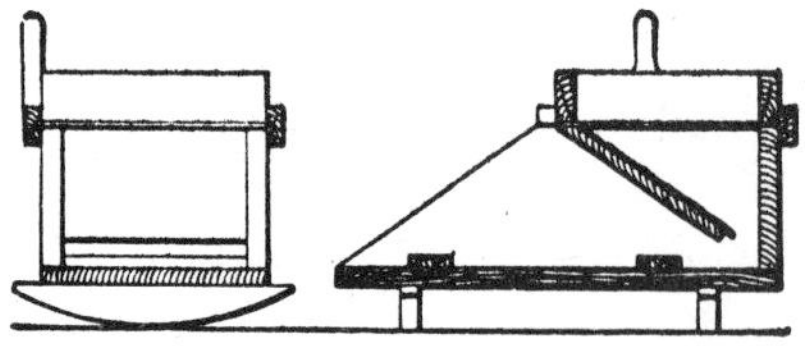

ABOVE: Rocker or cradle

RIGHT: Rocker operation

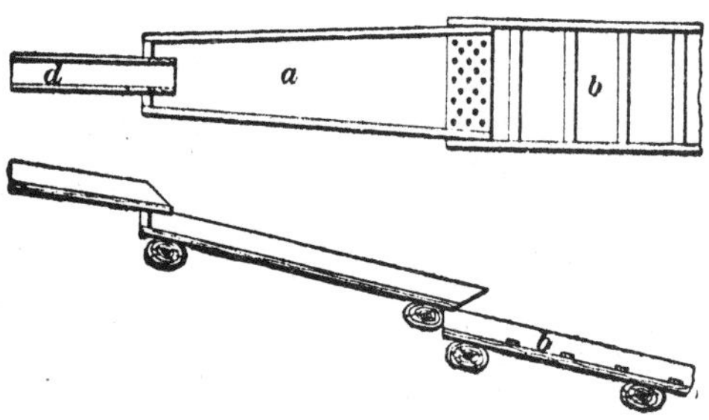

LEFT: Long Tom

BELOW: Broad Tom operation

The rest of our company has not made
anything. Duesler is at hose making.
We 16 This is one of those beautefull
clear days that is so common
in this country all hands ar pro
specting & mining but me & Dues
ler he is sewing on 50 yds of can
vass to make hose & I to help him
& allso to making other fixurs
to set up our knew pipe & hose.
There has no gold got to day.
Th 17 It is not necessary to say
anything about the weather as
it could not be finer. Shaw &
others set up our old hose & pipe
down on his & Dueslers claim
& made a commencemet at pipe
ing. I & Duesler finishd the new
hose & expect to set up on tomorrow
at our old diggins. John & Roister
got gold 39$. Lawrance & Shults
is taking it out evry day.
I thought several times to day
of making a guess what day
in this month I would see my
wife right here. But before I
came to any conclution as to the
day I could not determine.
Cash paid for nails 75 cts for comp.[22]

Quicksilver (mercury) was often introduced into the mining process in order to recover fine gold. Quicksilver acted as a collecting agent because of its ability to attach itself to gold. When placed in a pan or sluice the fine gold amalgamated to the mercury rather than be washed out of the device. Subsequently, the mercury amalgam was removed from the device and the gold separated from the mercury by application of heat—heat caused the mercury to evaporate, leaving the fine gold.

Augst 55
Su 5 Well this is another sabbath. Ive
read 1 & 2 chapt of Colosians, it being
my place of reading. We next got
down our sand bottle & emtied in

> c.t. a pan pourd in quick silver & pa
> nd out the sand & got out the qui
> ck silver 17$ & we rockd out of there
> square box 5 1/2$ but stopd for the
> want of a better rocker. Shaw cam
> e up to see if we was ready to let
> him off so we all went down to the
> point...[23]

Mining technology advanced rapidly. From the hand-operated devices operated by individual miners and partnerships which had worked the placer deposits near Nelson Point during the years of 1850-70, mining operations expanded to include large-scale enterprises which utilized expensive machinery requiring substantial investment. Initially, working placer deposits required little capital or experience, but the time soon came when gold was mined from shafts, tunnels, quartz veins, and by extensive hydraulic operations.

WINTER OPERATIONS.

Many miners returned to the lower valleys during the winter months of December, January, and February. However, a surprising number remained in the mountains, mining as often as the weather permitted. Haun, for example, remained at his claim year-round, describing the conditions this way:

> Feb 54
> Mo 6 Rained and snowed last knight
> mixd & raind all day. We went to
> washing out gold $11.25. This is
> the first since middle Dec. last
> and not bad for a start. I made
> a cape out of the legs of gum boots
> that kept my shoulders dry all day.[24]

FATALITIES.

Mining was a dangerous endeavor. Men drowned in the swift waters of the Middle Fork. Hydraulicing added new dimensions of danger. Haun noted one of the many local fatalities:

> July 54
> Sat 22 I went down to the point
> & got a letter for I Lloyd
> from his wife & was told the

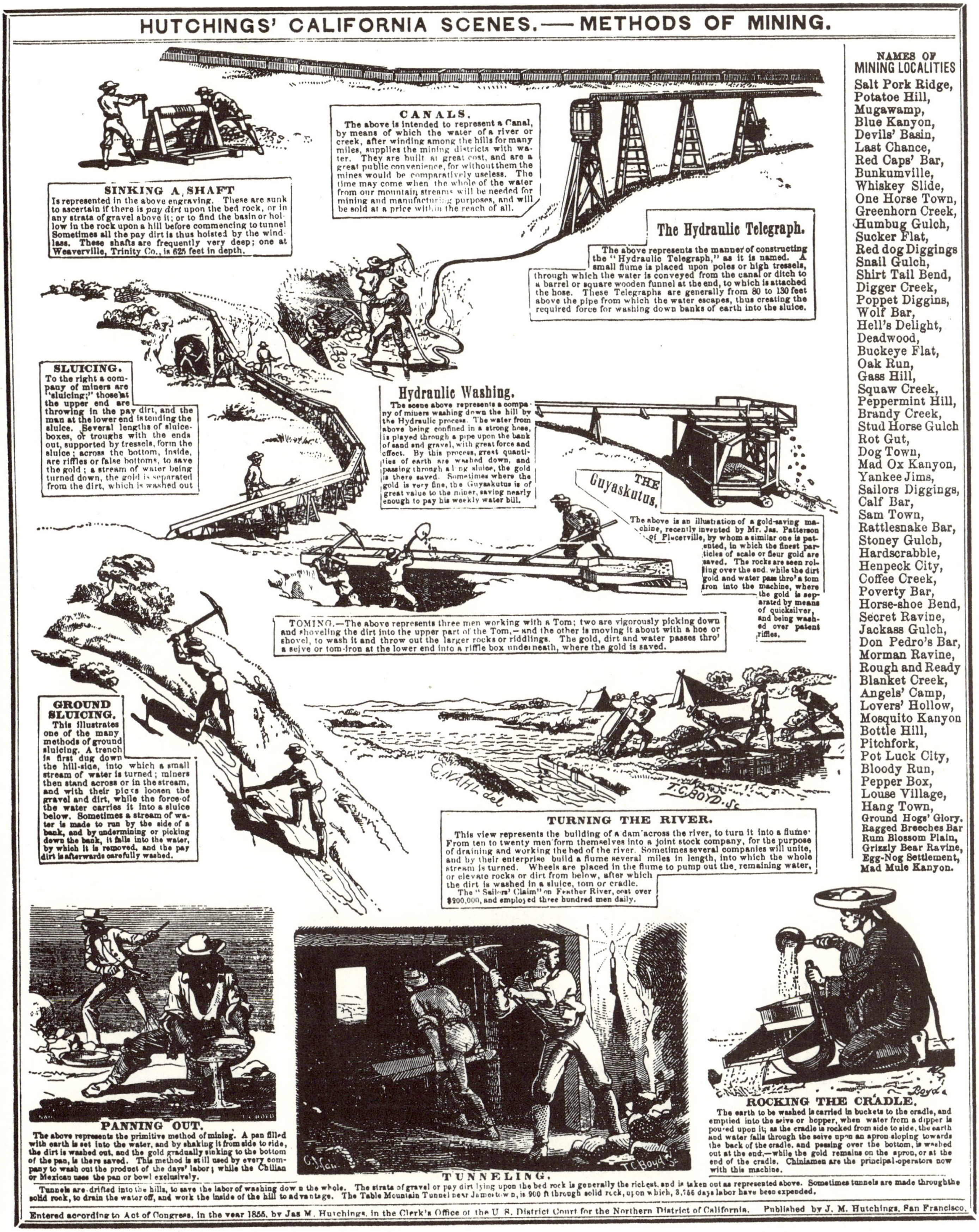

HUTCHINGS' CALIFORNIA SCENES.—METHODS OF MINING.

SINKING A SHAFT
Is represented in the above engraving. These are sunk to ascertain if there is *pay dirt* upon the bed rock, or in any strata of gravel above it; or to find the basin or hollow in the rock upon a hill before commencing to tunnel Sometimes all the pay dirt is thus hoisted by the windlass. These shafts are frequently very deep; one at Weaverville, Trinity Co., is 625 feet in depth.

CANALS.
The above is intended to represent a Canal, by means of which the water of a river or creek, after winding among the hills for many miles, supplies the mining districts with water. They are built at great cost, and are a great public convenience, for without them the mines would be comparatively useless. The time may come when the whole of the water from our mountain streams will be needed for mining and manufacturing purposes, and will be sold at a price within the reach of all.

The Hydraulic Telegraph.
The above represents the manner of constructing the "Hydraulic Telegraph," as it is named. A small flume is placed upon poles or high tressels, through which the water is conveyed from the canal or ditch to a barrel or square wooden funnel at the end, to which is attached the hose. These Telegraphs are generally from 80 to 130 feet above the pipe from which the water escapes, thus creating the required force for washing down banks of earth into the sluice.

SLUICING.
To the right a company of miners are "sluicing;" those at the upper end are throwing in the pay dirt, and the man at the lower end is tending the sluice. Several lengths of sluice-boxes, or troughs with the ends out, supported by tressels, form the sluice; across the bottom, inside, are riffles or false bottoms, to save the gold; a stream of water being turned down, the gold is separated from the dirt, which is washed out

Hydraulic Washing.
The scene above represents a company of miners washing down the hill by the Hydraulic process. The water from above being confined in a strong hose, is played through a pipe upon the bank of sand and gravel, with great force and effect. By this process, great quantities of earth are washed down, and passing through a long sluice, the gold is there saved. Sometimes where the gold is very fine, the Guyaskutus is of great value to the miner, saving nearly enough to pay his weekly water bill.

THE Guyaskutus,
The above is an illustration of a gold-saving machine, recently invented by Mr. Jas. Patterson of Placerville, by whom a similar one is patented, in which the finest particles of scale or flour gold are saved. The rocks are seen rolling over the end, while the dirt gold and water pass thro' a tom iron into the machine, where the gold is separated by means of quicksilver, and being washed over patent riffles.

TOMING.—The above represents three men working with a Tom; two are vigorously picking down and shoveling the dirt into the upper part of the Tom,—and the other is moving it about with a hoe or shovel, to wash it and throw out the larger rocks or riddlings. The gold, dirt and water passes thro' a seive or tom-iron at the lower end into a riffle box underneath, where the gold is saved.

GROUND SLUICING.
This illustrates one of the many methods of ground sluicing. A trench is first dug down the hill-side, into which a small stream of water is turned; miners then stand across or in the stream, and with their picks loosen the gravel and dirt, while the force of the water carries it into a sluice below. Sometimes a stream of water is made to run by the side of a bank, and by undermining or picking down the bank, it falls into the water, by which it is removed, and the pay dirt is afterwards carefully washed.

C. NAHL del

T. C. BOYD. Sc

TURNING THE RIVER.
This view represents the building of a dam across the river, to turn it into a flume. From ten to twenty men form themselves into a joint stock company, for the purpose of draining and working the bed of the river. Sometimes several companies will unite, and by their enterprise build a flume several miles in length, into which the whole stream is turned. Wheels are placed in the flume to pump out the remaining water, or elevate rocks or dirt from below, after which the dirt is washed in a sluice, tom or cradle.
The "Sailors' Claim" on Feather River, cost over $200,000, and employed three hundred men daily.

NAMES OF MINING LOCALITIES
Salt Pork Ridge,
Potatoe Hill,
Mugawamp,
Blue Kanyon,
Devils' Basin,
Last Chance,
Red Caps' Bar,
Bunkumville,
Whiskey Slide,
One Horse Town,
Greenhorn Creek,
Humbug Gulch,
Sucker Flat,
Red dog Diggings
Snail Gulch,
Shirt Tail Bend,
Digger Creek,
Poppet Diggins,
Wolf Bar,
Hell's Delight,
Deadwood,
Buckeye Flat,
Oak Run,
Gass Hill,
Squaw Creek,
Peppermint Hill,
Brandy Creek,
Stud Horse Gulch
Rot Gut,
Dog Town,
Mad Ox Kanyon,
Yankee Jims,
Sailors Diggings,
Calf Bar,
Sam Town,
Rattlesnake Bar,
Stoney Gulch,
Hardscrabble,
Henpeck City,
Coffee Creek,
Poverty Bar,
Horse-shoe Bend,
Secret Ravine,
Jackass Gulch,
Don Pedro's Bar,
Morman Ravine,
Rough and Ready
Blanket Creek,
Angels' Camp,
Lovers' Hollow,
Mosquito Kanyon
Bottle Hill,
Pitchfork,
Pot Luck City,
Bloody Run,
Pepper Box,
Louse Village,
Hang Town,
Ground Hogs' Glory,
Ragged Breeches Bar
Rum Blossom Plain,
Grizzly Bear Ravine,
Egg-Nog Settlement,
Mad Mule Kanyon.

PANNING OUT.
The above represents the primitive method of mining. A pan filled with earth is set into the water, and by shaking it from side to side, the dirt is washed out, and the gold gradually sinking to the bottom of the pan, is there saved. This method is still used by every company to wash out the product of the days' labor; while the Chilian or Mexican uses the pan or bowl exclusively.

TUNNELING.
Tunnels are drifted into the hills, to save the labor of washing down the whole. The strata of gravel or pay dirt lying upon the bed rock is generally the richest, and is taken out as represented above. Sometimes tunnels are made through the solid rock, to drain the water off, and work the inside of the hill to advantage. The Table Mountain Tunnel near Jamestown, is 900 ft through solid rock, upon which, 3,756 days labor have been expended.

ROCKING THE CRADLE.
The earth to be washed is carried in buckets to the cradle, and emptied into the seive or hopper, when water from a dipper is poured upon it; as the cradle is rocked from side to side, the earth and water falls through the seive upon an apron sloping towards the back of the cradle, and passing over the bottom, is washed out at the end;—while the gold remains on the apron, or at the end of the cradle. Chiniamen are the principal operators now with this machine.

Entered according to Act of Congress, in the year 1855, by Jas M. Hutchings, in the Clerk's Office of the U. S. District Court for the Northern District of California. Published by J. M. Hutchings, San Francisco.

Hutchings Magazine's "Hutchings' California Scenes—Methods of Mining."

> bank cavd in on a man by
> the of Douglap & 2 others
> on yesterday but the other 2
> got out & D was drowned
> in the mud...[25]

Edwards recorded the same incident: "On Railroad Flat Charles Douglas was killed and two others partially buried by the caving of the bank."[26] He also noted the death of Benomy Dick who was killed when a bank caved in on Fiddler's Flat.[27]

The diggings near Nelson Point provided some of the richest finds of the mother lode. Unfortunately, the actual locations of many of the most fertile sites have been lost to history. Of the most important diggings on Nelson Creek, Graveyard Flat, Fiddlers Flat, Buckeye Bar, Henpeck Flat, Grizzly Flat, Scotch Flat, Meeker Flat, and Independence Bar, few are recognized today.[28] As for locations all over the mother lode, Nelson Creek was discovered, inhabited and abandoned all on the basis of gold—its very existence ultimately depended upon the production of the diggings.

NOTES.

1 *Alta California*, Jly. 14, 1851, contained in *Bancroft Notes on Mines and Mining* (Bancroft Library: uncatalogued files).

2 Frank Lecouvreur, *From East Prussia to the Golden Gate* (New York: 1906), p. 243.

3 Charles D. Ferguson, *The Experiences of a Forty-niner During a Third of a Century in the Gold Fields* (Chico: 1924), pp. 81-82.

4 Kimball Webster, *The Gold Seekers of '49* (Manchester, N.H.: 1917), p. 157.

5 John K.Kepner, *Diary of John Kline Kepner* (1985), 1852.

6 Appendix III offers additional published accounts of Nelson Point mining activities.

7 See Appendix V.

8 Edwards, *Diary* (Bancroft Library: unpublished), vol. 4, Oct. 31, 1853.

9 Haun, *Diary* (Plumas County Museum Archives: unpublished), Jn. 18-22, 1854.

10 Ibid., May 28-29, 1855.

11 Other examples may be found in Appendix III.

12 Fariss and Smith, *History* (San Francisco: 1882), p. 288.

13 Edwards *Diary*, vol. 10, Aug. 13, 1855.

14 Hambly, *Autobiography* (Bancroft Library: unpublished), December, 1852.

15 Haun *Diary*.

16 *Quincy Union* 3, no. 34 (Jn. 17, 1865): 2.

17 State Mining Bureau, *Register of Mines and Minerals (Sacramento: 1898)*. See Appendix II for the text of the 1898 *Register of Mines and Minerals*.

18 *Plumas National Bulletin*, May 19, 1877.

[19] Ferguson *Experiences of a Forty-niner*, pp. 77-78.
[20] Haun *Diary*.
[21] Edwards *Diary*, vol. 6, May 4, 1854.
[22] Haun *Diary*.
[23] Ibid., Aug. 5, 1855.
[24] Ibid., Feb. 6, 1854.
[25] Ibid., Jly. 22, 1854.
[26] Edwards *Diary*, vol. 6, Jly. 21, 1854.
[27] Ibid., Aug. 19, 1854.
[28] For a complete list of Nelson Creek mining locations see Appendix I.

CHAPTER XXIII.

LUMBER.

...It is a branch, gathered from the prettiest feature of mountain scenery, a moss-grown fir-tree.

—Dame Shirley

The lumber industry developed out of the need to supply building materials to the mines. Lumber was required for the construction of flumes, sluices, wing-dams, etc. Later, timbers were needed for tunnel mines. In the early days the miners whipsawed their own lumber. Whipsawing was tedious work in which few miners possessed previous experience, as illustrated by Haun's description of his first attempt:

Jan 1855
Tu 16 We all, 5 of us concluded to get out some lumber for sloos boxes. John went & got a whip saw, by that time the rest of us had a log cut hewn lind & up on the pit ready for sawing, but it was work none of us had done., so we all tride but I, & Lawrance could saw the best. P.M. I sent John down to whet the saw cost 1$ for comp.
Wen 17 I whitted the saw and it went, but my arms is verry sore. We have made 4 planks in two pieces of days sawing...
Th 18 ...We all had a good time making our arms sore from working Brays whip saw.—
Fr 19 ...We finished sawing one log A.M. P.M. 3 of us cut & hewd another. Put it on the pit & lind it, & moovd the pit near the log...[1]

Commercial saw mills provided lumber in quantities sufficient to allow the miners more time to work the mines. Early mills relied on whipsawing. Steam-

Miners whipsawing on the Pacific Coast

operated machinery replaced those primitive operations late in the century. Of the mills that operated at Nelson Point, three can be identified:

Davis & Co.—operated during the 1850's
Sherwin's Mill—operated by J.L.C. Sherwin at Independence Bar during the 1850's
Nelson Point Saw Mill Company—operated by W.M. Metcalf from the late 1800's until the early 1900's

Sherwin's Mill was the most prominent at Nelson Creek, Edwards making numerous references to it. For example:

> Spencer went to the Saw Mill made a set of riffles Lewis & myself packed lumber for sluice...[2]

In 1863 the *Alta* reported of Nelson Point, "A saw-mill, to cost 25,000 is being erected there."[3]

By the turn of the century, mining demand for lumber had declined, yet demand remained high due to new housing. In 1902 the *Plumas Independent* reported:

Courtesy Jann Garvis

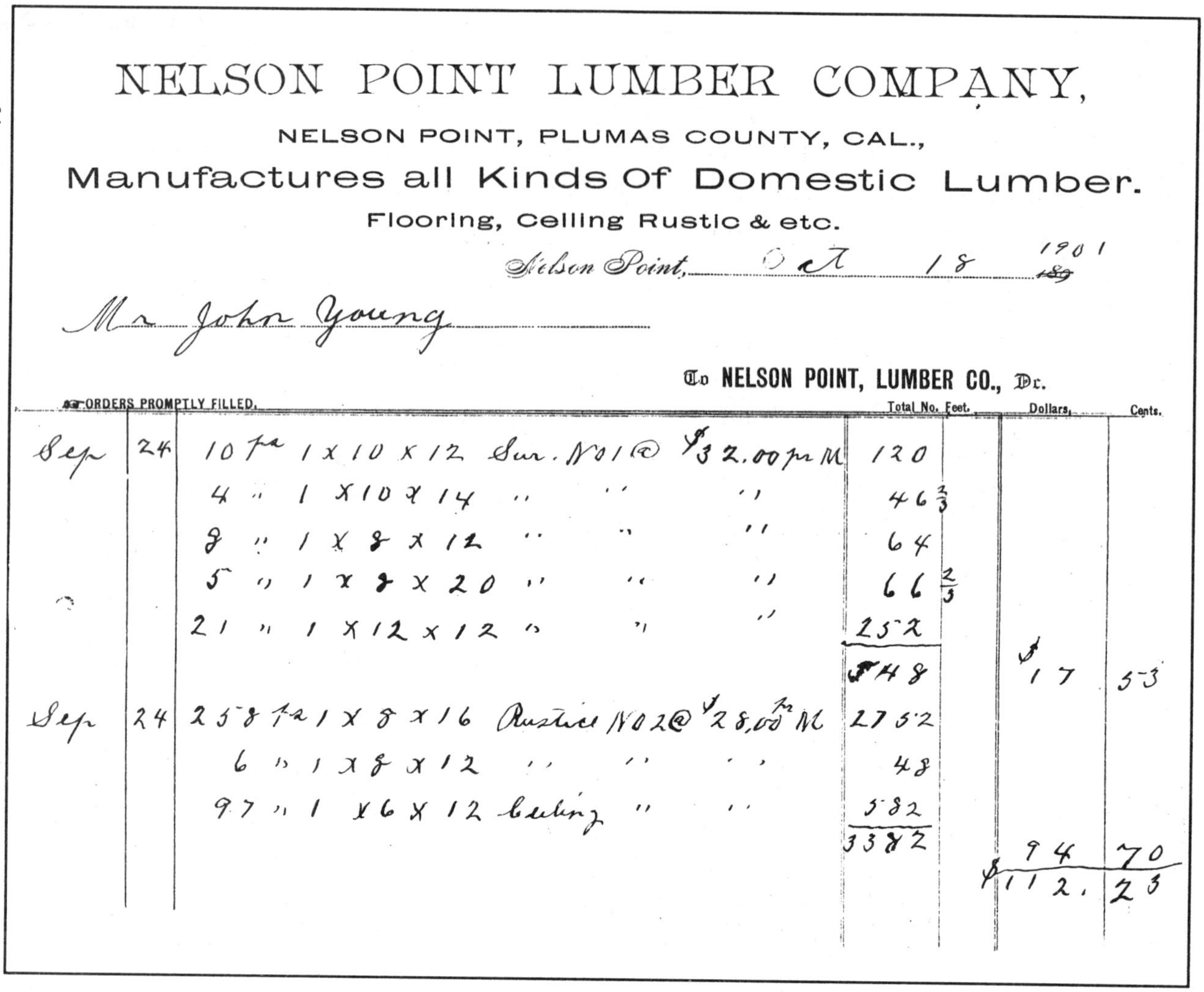

NELSON POINT LUMBER COMPANY,

NELSON POINT, PLUMAS COUNTY, CAL.,

Manufactures all Kinds Of Domestic Lumber.

Flooring, Ceiling Rustic & etc.

Nelson Point, Oct 18 1901

Mr John Young

To NELSON POINT, LUMBER CO., Dr.

ORDERS PROMPTLY FILLED.

			Total No. Feet.	Dollars.	Cents.
Sep	24	10 pa 1 x 10 x 12 Sur. No 1 @ $32.00 pr M	120		
		4 " 1 x 10 x 14 " " "	46 2/3		
		8 " 1 x 8 x 12 " " "	64		
		5 " 1 x 8 x 20 " " "	66 2/3		
		21 " 1 x 12 x 12 " " "	252		
			548	$17	53
Sep	24	258 pa 1 x 8 x 16 Rustic No 2 @ $28.00 pr M	2752		
		6 " 1 x 8 x 12 " " "	48		
		97 " 1 x 6 x 12 Ceiling " "	582		
			3382	94	70
				$112.	23

ABOVE AND OPPOSITE: Nelson Point Lumber Company receipts, circa 1900–1901

> Nelson Point Saw Mill Company. Lumber of all kinds and description to be had at short notice either in the yard or delivered in the Slate Creek Basin or American Valley from the company's mill at American House on the La Porte road. Saw mill dressed sugar and yellow pine a specialty. Rustic, flooring, wainscoting and finishing lumbers of A No.1 quality to be had is season. Rough lumber for building purposes furnished on short notice at prices to suit the times.
>
> W.M. Metcalf, Manager[4]

And in 1907:

> William Metcalf Jr. was in town from Nelson Tuesday and informed us that the mill would finish an order for 100,000 shingles this week.[5]

As mining activity declined, the need for lumber shifted from constructing flumes to building homes. Completion of the railroad enabled timber products to be readily shipped out of the county, thereby completing Plumas County's transition from a mining region to that of a center for lumber production.

NOTES.

1 Haun, *Diary* (Plumas County Museum Archives: unpublished), Jan. 16-19, 1855.

2 Edwards, *Diary* (Bancroft Library: unpublished), vol. 6, Apr. 3, 1854.

3 *Alta California*, Jan. 9, 1863.

4 *Plumas Independent*, Jan. 29, 1902.

5 Ibid. 15, no.2 (Sept. 25, 1907).

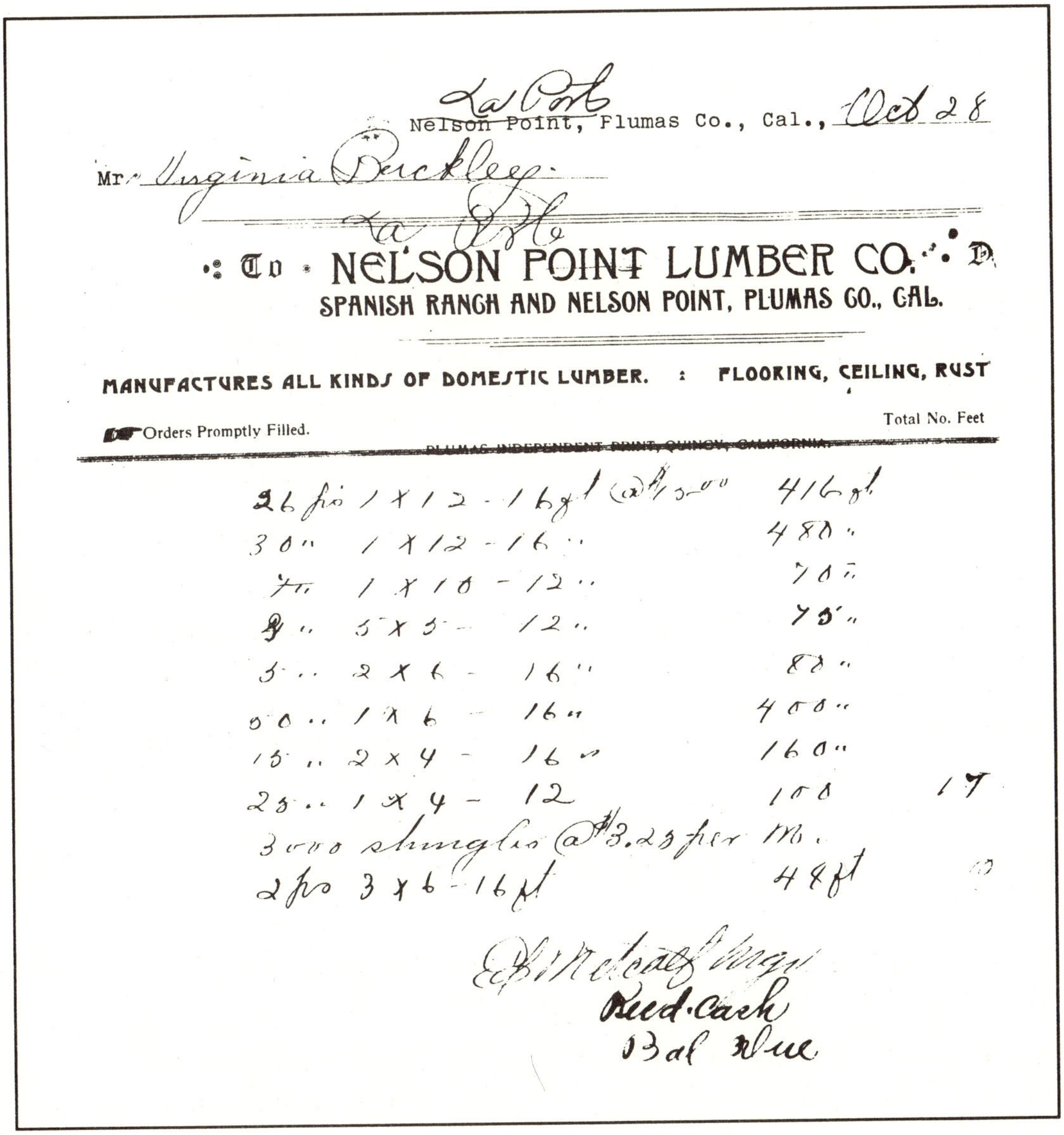

La Porte
~~Nelson Point~~, Plumas Co., Cal., Oct 28

Mr. Virginia Buckley
La Porte

To NELSON POINT LUMBER CO. Dr.
SPANISH RANCH AND NELSON POINT, PLUMAS CO., CAL.

MANUFACTURES ALL KINDS OF DOMESTIC LUMBER. : FLOORING, CEILING, RUST

Orders Promptly Filled. Total No. Feet

PLUMAS INDEPENDENT PRINT, QUINCY, CALIFORNIA.

26 pcs 1 x 12 - 16 ft @ $15.00	416 ft
30 " 1 x 12 - 16 "	480 "
7 " 1 x 10 - 12 "	70 "
9 " 5 x 5 - 12 "	75 "
5 " 2 x 6 - 16 "	80 "
50 " 1 x 6 - 16 "	400 "
15 " 2 x 4 - 16 "	160 "
25 " 1 x 4 - 12	100
3000 shingles @ $3.25 per M.	
2 pcs 3 x 6 - 16 ft	48 ft

E. F. McNeal, Mgr.
Recd. Cash
Bal Due

Courtesy Jann Garvis

CHAPTER XXIV.

THE WINTER OF '53.

The whole Bar is thickly peppered with empty bottles, oyster cans, sardine boxes, and brandied fruit jars, the harsher outlines of which are softened off by the thinnest possible coating of radiant snow.

—Dame Shirley

Three disasters devastated Nelson Point: the winter of 1853, the floods of 1862, and the fire of 1924. The winter of 1852-53 resulted in death and destruction throughout the northern lode. Mountain trails became blocked with snow in November, catching the miners short of supplies. Camps were deserted as the miners rushed to all parts of the county to secure supplies. Many miners from Nelson Point spent the winter in Onion or Strawberry Valley, where supplies were brought from Marysville.[1]

By near superhuman efforts, pack trains were brought in as close as possible to Nelson Point. The packers carried the supplies the remainder of the distance on their own backs. In one instance recorded, a man took a pack train to within one-half mile of his destination, and then had to pay men fifty cents a pound to have the goods taken the rest of the distance. At another point, a train of mules refused to go over the top of the mountain where the wind had piled the snow into a huge bluff. So they were blindfolded, led to the top, and pushed over, rolling with their packs clear to the bottom. Apparently, none of the mules was harmed. The miners resorted to extreme measures such as these so that food could be taken to those who might otherwise have starved.[2]

Remaining in the mountains too long often carried deadly consequences:

> Early in the month of January another party of twelve started from Nelson Point to make their way out of the mountains. The first day they reached Onion Valley, where they remained for the night. In the morning, six of them, led by J. H. Whitlock, started to break a path through the snow, it being agreed that the other six would follow their trail at noon and break the road for the balance of the day. The first party became lost in the snow, and wandered about for three days, suffering from the cold and hunger to a degree beyond description, and finally reaching a house alive, but in a most pitiable condition. The second party followed the trail of the first until they, too, became bewildered, and sought to retrace their steps to Onion valley. One of them succeeded in doing this, and sent help to the others. The relief party found Walter Goodspeed dead, and H. Brown and William Phillips so badly frozen that they both died afterwards.[3]

Hambly recorded the same incident:

> ...Whitlock returned [to Fiddler's Flat] and took possession and Goodspeeds interest in these claims were sold, who was frozen to death at Onion Val last Dec on his way down below to Marysville. My brother and several others got through all right. all had a close call...[4]

In the spring of 1853 the miners returned to their claims, life returned to normal, and the winter of '53 became a topic for conversation around campfires for years to come.

NOTES.

1 Fariss and Smith, *History* (San Francisco: 1882), pp. 152-53.

2 Ibid., p. 153.

3 Ibid., p. 155.

4 Hambly, *Autobiography* (Bancroft Library: unpublished), January, 1853.

CHAPTER XXV.

THE FLOODS OF '62.

How the haughty river seemed to laugh to scorn the feable efforts of man!
—Dame Shirley

Building Nelson Point on its original site was an exercise in futility. The town originally located in a rugged ravine nearly straight up and down on either side. Setting a firm building foundation must have been extremely difficult, if not impossible. When the rains came in January of 1862, it is not surprising that Nelson Point went down the river—literally! Two excellent accounts exist of the devastating events which began on Friday the 10th of January:

> About 5 o'clock on Friday morning last, the Feather River sent a huge log into the rear of Cunningham's stor, carrying away a shed that was attached to the main building with such a tremendous crash as to awaken the sleeping inmates of the store in double quick time. This was the first greeting of the freshet at Nelson.
>
> The day dawned on a crowd of very busy men packing the contents of the store to a place of safety. Myers, the proprietor of the saloon, now deemed it expedient to remove the wares from the exposed tenement, and a busy scene struck up on that side of the town also.
>
> The restraurant owned by Myers & Richards was next cleaned out of all its movables and by 12 o'clock noon, all hands stood watching to see those buildings swept away.
>
> The new bridge across Nelson Creek now rose at one end, swung against the side of the saloon, making a tremendous crash, and floated off.
>
> At 4 o'clock on Saturday morning, the waters of Feather and Nelson were at the culminating point. Cunningham & Co.'s store now rose, swung gradually around, and went gliding rapidly down the river for several hundred feet. It then turned into an eddy in the side of the stream, came back close to where it had started from; the boys gave three hearty cheers; the store then shot out into the middle of the river and was soon out of sight.
>
> A cheerful fire was burning in the stove; the lights from the candles shown brightly out of the doors and windows, and a faithful watch-dog stood at the open door, looking out as composed, as if nothing unusual had happened. In this manner, the store passed around a bend in the river, and was soon out of sight.
>
> The saloon remained a few minutes longer and then prepared to take its departure.
>
> It had been lighted up on purpose to see it take its leave, and a more graceful, pretty scene could not be imagined. It came back in the eddy just precisely as the store had done and then shot out into the middle of the river, amid cheers of the spectators and then glided away.

The losses I have heard estimated as follows: Cunningham & Company, $2,000; J. Richards $500, and I expect these figuers are pretty nearly correct.

The lively times that will follow, will make those who have been damaged by the freshet soon forgot their losses. A freshet like that of '52 has long been wished for by the Nelsonites, and a greater than that of '52 has answered to their call.

The tailings and all the dams have been swept out of the beds of the streams leaving an excellent opportunity for fluming, the coming summer. Very many of the bars have been carried up in front too deeply to pay for working, but since they have been subjected to the sweeping process, will pay good wages.

The Chinese have been notified to leave instanter.[1]

Nelson Point Corresponence.

Nelson Point, Jan. 15th, '62.

Friend Lynch:—As it is now some three weeks since we have been blessed with the sight of a STANDARD, it will be well enough to send you this for the double purpose of reminding you that there was such a place as Nelson Point, and, also, to let you know that it has gone down—the river.

On Friday last we were visited by the most destructive and devastating flood that it has ever been the lot of 'white' men to see in this part of the country. Feather river reached a hight of nine feet more than than was ever known by the 'oldest inhabitant,' carrying away bridges, camps, store, saloon, restaurant, and much real estate.—But to particularize: Friday morning, Jan. 10th, the water had risen nearly to the mark of the last high freshet, and it still continued raining very hard. We removed the goods from the store of Capt. Cunningham & Co., and about noon, the drift-wood had wedged itself under it, so as to lift it from its foundation, and it started for the lower country—not having the Captain on board, and the pilot was either drunk or absent from his post. In rounding the point of rock below the town, it struck, and instantly vanished from our sight. Next, the stable and butcher shop, belonging to N. Pyles followed suit,; then the Nelson Creek bridge—a new one just completed—took up its line of march; also, the timber of H. White for the Feather bridge, which was nearly ready to raise; and about daylight, Saturday morning; the saloon belonging to Mr. Myers deliberately sailed off as though it had fully made up its mind to follow its companions; the restaurant building followed soon after—everything showing a downward tendency. The blacksmith shop, being an old settler, and not ready to leave, careened over to an angle of forty-five degrees, shoreward, and so saved itself. About Saturday, noon, the water had reached its highest mark,, which is some twenty-eight feet above its ordinary level. Logs and drift are now lying up in places that, when the water falls, will cause men to wonder with the poet:

"Not that the thing was rich or rare, but how the devil it got there."

The Sherwin mill dam has gone; also, the dam of Keen, Stevens & Co., and some five hundred feet of their flume. White & B o, saved their camp, this time, by working all night; the water ran in at the windows, but with the assistance of Providence, about twenty tons of rock and a strong guy rope, they are now safe. All the Rocky Bar boys are washed out—their camps all gone; but they are in no way discouraged, as since the flood they have picked up nearly enough gold from the rocks, that have been laid bare by the late freshet, to build a fine two story brick.

Our creek and river are well cleand out, the strong current having swept away the tailings, and uncovered to the bed-rock most of the low bars. There are many places that will now pay us rich that have been covered up since '52 and working them will be easy and profitable. We hope to reap such compensation from them as will repay us for the damage done by the flood. I prophecy very lively times here the coming season—there will be more money taken out than ever, and we will all, without a doubt, play even.

Yours,

PLUG.[2]

The flooding of '62 was not restricted to Nelson Creek. All over the northern lode the effects of flood waters were similar:

Mining in California during 1862.
The Floods of the Winter of 1861-2.

The immediate damage to the miners from the floods of the winter of 1861-2 was immense. It stopped mining operations of nearly all kinds in many districts for a month or six-weeks, throwing large numbers out of employment, while at the same time, by sweeping away the bridges and rendering the roads impassible, it cut off their supplies of provisions and greatly enhanced the cost of living...On the other hand, the immense quantity of water which fell was not without its compensating benefits. Many of the smaller gulches and higher flats were thus supplied with sufficient water to permit their being worked with great success by the miners thrown out of employment elsewhere. Besides this, the beds of many of the rivers, creeks, and ravines, which had become filled with worthless tailings from hydraulic and other claims, were suddenly cleared out, often to a depth of ten feet or more, so that the bed rock could again be reached and worked with profit...[3]

The floods of '62 both devastated and benefitted Nelson Point. However, improved mining in some areas came at a high price. Had the buildings been more securely located, the extent of the damage would have been greatly reduced.

NOTES.

1 *Plumas Standard*, Feb. 8, 1862, contained in Ruth Clary's Scrapbook, Plumas County Museum Archives.

2 *Plumas Standard* 3, no. 16 (Feb. 1, 1862): 2.

3 Hubert H. Bancroft, *Bancroft Notes on Mines and Mining* (Bancroft Library: uncatalogued), p. 332.

CHAPTER XXVI.

THE FIRE OF 1924.

The mere handful of miners still remaining had seen the town spring up, spread, grow, and flourish in its pride; and they had seen it sicken and die, and pass away like a dream.
—Mark Twain

Pauly's Hotel stood as the final landmark of Nelson Point until 1924. Then the last of the three catastrophes appeared over the treetops—a devastating forest fire. The fire began at the Nelson Creek bridge. Initially it appeared that Nelson Point would be spared. While Thursday's paper reported:

Careless Campers Cause Destructive Nelson Point Fire

For a time Wednesday afternoon the Nelson Point hotel was in extreme danger, the structure being saved from the flames only by backfiring and constant wetting by bucket brigade. The hotel is owned and operated by Mrs. Kate Pauly who, with O. Moncur, father of J.O. Moncur of Quincy, were the only occupants when the fire started.[1]

The following edition of the *Bulletin* told the entire story:

Nelson Point Fire Destroys Landmark of Early Gold Rush

Hotel Erected in Early 60's Goes Down Before Flames—
Owned and Operated By Pauly Family Since 1871—
Contents Total Loss and Proprietress Barely Escapes.

Another landmark of the early pioneer days, the Nelson Point Hotel, 12 miles from Quincy on the La Porte road, has passed into history. The fire that started at the Nelson Creek bridge, the morning of July 30, reached the old hotel about noon Thursday and in less than an hour there was nothing left on the grounds to indicate that it had ever been a human habitation.

The spectacle at Nelson Point, according to B.F. Pauly, of Quincy, who had hastened to the scene to look after his mother, Mrs. Kate Pauly, proprietress of the hotel, was terrible, yet magnificent.

Pauly says the flames would jump a half mile when some cinder carried by the wind would light in dry grass or pine needles. The men fighting the fire were helpless to do anything and in some cases it required quick action to avoid being enmeshed by the flames and burned to death.

Wednesday evening and Thursday morning, Pauly says, the hotel was apparently in no serious danger; then a change in the wind swept the flames rapidly to the hotel. Mrs. Pauly had made preparations to evacuate the building but seeing the fire receding began

> to think the structure was safe. The sudden change in the wind required quick action to get Mrs. Pauly out of the hotel and but little of the contents of the building were saved. A couple of trunks and a few trinkets were all that could be secured, Mrs. Pauly even lost about $100 in money, as there was not enough time to open the safe...
>
> The burning of the Nelson Point Hotel adds one more to the list of passing sign posts, which includes the Plumas House at Quincy, the Meadow Valley Inn, The Greenville Hotel, the Sierra Hotel at loyalton, and the Johnsville Hotel.
>
> During the progress of the fire in the vicinity of Nelson Point four automobiles were destroyed, including a big Hudson belonging to J.P. Crumpton, a mining man; also Messrs Whitney. Hughs and James each lost an automobile. These are all mining men who were working in the neighborhood.[2]

The fire swept through Nelson Point in a northeasterly direction burning towards Sloat. Simply stated, it erased Nelson Point from the map.

NOTES.

1 *Plumas National Bulletin* 47, no. 47 (1924): 1.
2 Ibid. 48: 1.

CHAPTER XXVII.

CONCLUSION.

The burning of the Pauly Hotel marked the final chapter of Nelson Point. However, Nelson Point was already a shadow of its past by the time the hotel burned—a mere reminder of what once had been. Its significance had already been reduced to a traveler's inn and supply center for a few fizzling mines and the handful of diehard miners who worked them. Nonetheless, its discovery, boom, decline, and ultimate destruction over three quarters of a century paints a familiar portrait of the evolution of countless gold rush camps that vanished without a trace.

Nelson Point became one of but a handful of camps in the region to evolve into a town. Why then did it not survive as did nearby La Porte or Quincy? Topography was the single most limiting factor. Even though La Porte was situated deeper in the Sierra Nevada, the town was located on relatively level land which supported the growth of a significant town center. Nelson Point's deep and narrow ravine proved to be susceptible to flooding and ultimately inappropriate for expansion. By the time Nelson Point was relocated to Lindsay Point, its placer deposits had played out and its population had dwindled to a fraction of former count. Consequently, plans to revive Nelson Point commerce on the La Porte Road experienced limited success, which gradually faded as the lode mining operations petered out. On the other hand, La Porte's lode mining operations proved to be far more extensive and long lasting. But even in La Porte the mining operations continued to support only a relatively small population. Only Quincy survived the fading gold rush era in the region as a significant community.

The foremost reason for Quincy's growth during the gold rush was its selection as the county seat. Though nearby Elizabethtown was seriously considered because it too was centrally located, like Nelson Point it was located in a ravine. Neither town could compete with Quincy's location in the American Valley. Quincy developed as the center of the support system for the surrounding mining community and flourished during the gold rush due to its political and commercial significance. However, Quincy prospered after the gold rush because it offered what Nelson Point could not—a place to raise a family.

During the early years of the rush the American Valley became a refuge from the harsh conditions of diggings such as Nelson Point. Places like the American and New England Ranches offered miners a tranquil break from the primitive conditions and animation of Nelson Point. Consequently, it is not surprising that, when joined by their wives and faced with the decision of whether to return east or to begin a new life in California, often they settled in the American Valley. There gold purchased fertile land where a couple could raise not only farm products but a family.

Nelson Point offered only rugged living conditions which were seldom tolerated for any length of time by families. And the vicinity offered little opportunity for the kind of family settlement experienced by the Paulys. So when the gold played out at surviving gold camps throughout the Sierra Nevada such as Nelson Point, all that remained was timber. Plumas County became timber country. However, the logging industry supported but a fraction of the population that gold had attracted. Consequently, most Nelsonites moved on in search of fortune and adventure leaving Nelson Point like so many gold camps throughout the northern mother lode as lost memories of the gold rush era.

Nelson Point played a significant role in the development of Plumas County. Haun, Thompson and Whitlock were but a few of the many Nelsonites who were instrumental in developing the infrastructure that survived the gold rush and established Quincy as the hub of Plumas County. From a broader perspective, Nelson Point was one of hundreds of now-forgotten gold camps that played an essential role in the establishment of California and subsequent settlement of the far west. With each swing of the pick, Nelsonites added a stroke on the canvas—a canvas that would eventually convey a portrait of the California gold rush.

APPENDICES.

APPENDIX I.

MINING LOCATIONS.

Contained herein are the principal mining locations in the vicinity of Nelson Point. Sites are located on Nelson Creek unless otherwise noted. Spellings are as they appeared in original citations.

Baker's Flat...on the Middle Fork below Rich Bar
Big Flat...on the Middle Fork
Bogus Flatt...above Sterling Flat
Brays Bar...between Henpeck and Sterling Flat
Braz's Bar
Buckeye bar
Clouneys Flatt
Columbia Flat...on the Middle Fork downstream from the confluence of Nelson Creek
Dixon creek...feeds into Nelson Creek at Independence Bar
Fiddlers flat
French Hill
Frenchman's Bar
Graveyard flat
Grizzly flat
Henpeck flat
Hospital Flat
Independence Bar...at the confluence of Nelson and Dixon Creeks
Joes ravine
Maddens Flat
Maddens Ravine
Masquito Bar
Meeker flat
Missouri Falls...on the Middle Fork below Rocky Bar
Mountain Flatt
Picayune Bar
Railroad Flatt
Railroad Point
Ranch Bar
Rich Bar...on the Middle Fork below Missouri Falls
Rocky Bar...on the Middle Fork below Columbia Flat
Sailor Flatt
Scotch flat
Scotchmans Bar
Syracutus Bar
Walkers Flatt
Whitney's Bar
Wileys Flatt

APPENDIX II.

1898 REGISTER OF MINES AND MINERALS.

Contained herein are excerpts from the state 1898 Register of Mines and Minerals that pertain to Nelson Point.

REGISTER OF MINES AND MINERALS

COUNTY OF PLUMAS

CALIFORNIA

ISSUED BY THE

STATE MINING BUREAU

A. S. COOPER
STATE MINERALOGIST

Data Collected October, 1898

THE Register of Mines and Minerals of the State of California, issued by the State Mining Bureau, is intended to give information concerning the gold mines—quartz, hydraulic, drift, placer, etc.—as well as a list of mills and arrastras; and also to give corresponding information relating to each of the various mineral substances mined or commercially utilized in the State.

This forms practically both a State and County directory of the mines of California, the Register being printed so that each county is represented in a separate pamphlet. The Register will answer for the guidance of those who want to look at any particular mine, or to learn the essential facts concerning it.

Of course some names may be omitted owing to the reluctance or indifference of the owner or superintendent in furnishing the necessary data, or by being overlooked by the Field Assistant of the Bureau.

It is intended, however, that in time such omissions will be filled in and in that event a supplement will be published and attached to the Register of each county. Such supplements will contain not only corrections to date, but the names of all new properties and owners.

It has been the intention of the State Mineralogist to obtain the desired information concerning the mines and minerals and to publish only the essential facts in as condensed a shape as possible. For this reason the tabular form has been adopted, so as to reduce the verbiage necessary in a printed report of the ordinary character, where the essential description, slightly varied, must be repeated in each paragraph.

Those things only are published which are of general interest and which the mine-owner has no reason to object making public. It is not considered the province of the Bureau to "expert" mining properties or to pass on their standing or value one way or another. Should more detailed information be desired by any individual, the addresses of owner and superintendent are given and they may be communicated with.

At the same time the figures given under "developments" will show at once whether the mine is a fully developed one or a prospect, so that any possible misrepresentation may be checked.

The information printed is of such a character that it cannot do any injury to the reputation of the mine described, but will be of service to the owners and to the public. In order to have the data as correct as possible, only actual residents of the respective counties, familiar with their own localities, have been employed as Field Assistants to prepare the Register.

To accompany the printed Register is a topographical map of each county showing all mineral locations, etc. These maps are on the large scale of two miles to one inch and produced in three colors. The mines are numbered in the table and a corresponding number on the map indicates the exact location. One may find from this approximately how far he must go from any specified town or station to reach any particular mine, and may learn the location of railroads, roads, trails, etc.

The object of this publication is to give in as condensed and convenient form as possible such information concerning the mines of California as is generally sought by intending investors or others interested, and to show the location of each mine in the respective counties.

Registers of all the counties of the State will be given to the public as soon as the work in each is completed and can be published.

It is earnestly requested, if any errors or omissions are noted, that the proper corrections will be sent at once to the State Mineralogist, State Mining Bureau, Union Ferry Building, San Francisco, Cal.

REGISTER OF MINES AND MINERALS

PLUMAS COUNTY.

	Name of Mine.	Nearest Town.	Location. Section.	Township.	Range.	Map No.	Whether Patented.	Elevation.	Area.	Name of Stream.	Length of Channel.
								Feet.	Acres.		
1	Alum Cove	Butte Valley	17	26	8	16		3300	20	North Fork Feather River	
2	Amigo Gold Channel	Laporte	10	21	8	229		5000			
3	Bald Mountain	"	3	21	9	233	yes	5100	320	Rabbit Creek	
4	Barker Hill	Butte Valley	8	26	8	10		4600		North Fork Feather River	
5	Beckwith		29	22	11	211		6500	320	Nelson Creek	
6	Bolyar	Spanish Ranch	2	24	8	86		4000		Spanish Creek	
7	Bull	Quincy	25	24	9	115		4500		Mill Creek	
8	Bunker Hill	Gibsonville, Sierra Co.	15, 16	22	10	192		6000		Hopkins Creek	
9	Bushman	Quincy	33, 34	25	9	73		3700		Blackhawk Creek	
10	California Klondike (a)	Buck P. O.	25	24	7	82		6500		Bucks Creek	
11	Cameron	Butte Valley	16	26	8	13		3300	60	North Fork Feather River	
12	Cash Entry	Chaparral, Butte Co.	24	25	5	40	yes	5000	150	Rock Creek	
13	Centennial	Johnsville	19	22	11	210		6000	90	Nelson Creek	
14	Chaparral Hill	Spanish Ranch	30	25	8	58		5100		North Fork Feather River	
15	Claremont Hill	Quincy	2, 3, 11	23	9	144		6600		Spanish Creek	
16	Cole's Diggings	Laporte	29	21	9	248		5000		Slate Creek	
17	Deadwood	Meadow Valley	29	24	8	93		4700		Big Creek	
18	Delaware (b)	Johnsville	17	22	11	205					
19	Dutch Hill	Butte Valley	8, 17	26	8	11		4692	80	North Fork Feather River	
20	Dutch Mine	Laporte	4	21	8	227	yes	4800		Fall River	
21	Eclipse	Eclipse P. O.	3	22	10	188	"	6500		Nelson Creek	
22	Edison	Chaparral, Butte Co.	24	25	5	39	"	5000		Rock Creek	
23	Elizabethtown Flat	Quincy	10	24	9	106	"	3500		Spanish Creek	
24	Emerson & McMullin	Genesee	3	24	11	121		5150	80	Grizzly Creek	
25	Emigrant Hill	Quincy	3	24	9	124		3900		Spanish Creek	
26	Enslow	Laporte	10	21	8	230	yes	4800		South Fork Feather River	
27	Erickson	Spanish Ranch	1	24	8	89		3800		Spanish Creek	
28	Eureka Consolidated (c)	Laporte	35	22	9	185	yes	5700		Slate Creek	
29	Excelsior (d)	"	9	21	9	239	"	5200	635	" "	
30	Fall River (e)	"	9	21	8	228	"	4800		Fall River	
31	Fortuna	"	31	22	9	178		5000		South Fork Feather River	
32	Franklin Consolidated (f)	Johnsville	20	22	11	209		6500	320	Nelson Creek	
33	Fred Scott	Butte Valley	17	26	8	14		3300	120	North Fork Feather River	
34	Glazier (and Glazier No. 2)	" "	9	26	8	9		3300	140, 30	" " " "	
35	Gloria Mundi	Quincy	26	24	9	116		4500		Spanish Creek	
36	Golden Ancient Channel	Spanish Ranch 2, 3, 10, 22, 11, 14,	15, 23	23	8	142		5000	1460	Bear Creek	
37	Golden Summit (g)	Chaparral, Butte Co.	21	25	5	38	yes	6200	320	North Fork Feather River	
38	Golden Wreath (h)	Quincy	17, 18	24	9	110	"	3400	160	Spanish Creek	
39	Great Republic (h)	"	29, 30	24	11	126	"	4200	160	Greenhorn Creek	
40	Grizzly (h)	"	30	24	11	127	"	4500	160	" "	
41	Helm	Laporte	32	22	8	176		5300	160	Middle Fork Feather River	
42	Kenzie Consolidated (i)	"	27	22	9	181	yes	5700	471	Slate Creek	
43	Last Chance (i)	"	2	21	9	234	yes	5700	594	" "	
44	Last Resort	Quincy	33	24	9	117		5500		Rock Creek	
45	Lava Bed	Spanish Ranch	2	24	8	87		4400		Spanish Creek	
46	Louise (j)	Laporte	24, 28	22	9	182	yes	5300	506	Slate Creek	
47	Lost Channel	Cromberg	19	23	12	169		4500	80	Middle Fork Feather River	
48	Malvern Hill	Butte Valley	18	26	8	17		3550	80	North Fork Feather River	
49	Meyer	Cromberg	20	23	12	170		4900	70	Middle Fork Feather River	
50	Miller	Johnsville	32	23	11	164		5500	138	" " " "	
51	Moraine	Cromberg	29	23	12	171		5500	160	" " " "	
52	Murdoch (Mendota Hill)	Spanish Ranch	1	24	8	88		3800		Spanish Creek	
53	Newtown Flat	Quincy	3	24	9	102		3700		" "	
54	Oliver	Eclipse P. O.	36	23	9	145	yes	5500	576	Onion Valley Creek	
55	Peter Henry	Quincy	8	23	10	147		4900	80	Willow Creek	
56	Pine Creek (k)	"	29	24	11	125	yes	4500	90		
57	Pittsburg and Pennsylvania (l)	Johnsville	5	22	11	195		5500	320	Nelson Creek	
58	Pliocene	Butte Valley	4, 9	26	8	7	yes	4050	100	North Fork Feather River	
59	Plumas Klondike	Quincy	7	23	10	148		4900	140	Willow Creek	
60	Post	Laporte	11	21	9	237		5400	90	Slate Creek	

[a] Relocation of the old Spanish Peak mine. [b] See Franklin Consolidated. [c] Worked in conjunction with other mines from the Thistle shaft, Sierra County. [d] See United Placer mine.
[k] See Grizzly and Republic mines. [l] See Miller mine.

REGISTER OF MINES AND MINERALS

PLUMAS COUNTY.

	Name of Mine.	Nearest Town.	Section.	Township.	Range.	Map No.	Whether Patented.	Elevation.	Area.	Name of Stream.	Length of Channel.
								Feet.	Acres.		
61	Quigley	Laporte	29	22	9	180		5000	160	Slate Creek	
62	Ray & Traynor	Cromberg	1	23	11	158		5000	180	L. Long Valley Creek	
63	Red Ravine	Johnsville	25	22	10	194		5600	130	Nelson Creek	
64	Red Rose	"	25	22	10	193		5600	160	" "	
65	Red Slide	Eclipse P. O.	4	22	10	187	yes	6500	50	" "	
66	Riverdale	Quincy	10	24	9	107		3450		Spanish Creek	
67	Rockwell	Spanish Ranch	17	25	6	41		4500	40	Chips Creek	
68	Salmon Falls	Butte Valley	4	26	8	5		4100	80	North Fork Feather River	
69	San José	" "	17	26	8	15		3500	170	" " " "	
70	St. Louis	Laporte	10	21	9	236		5400		Slate Creek	
71	Spring Hill	Meadow Valley	34	24	8	101		5000		Meadow Valley Creek	
72	Summit	Spanish Ranch	34	25	8	61		4750	320	Spanish Creek	
73	Sunnyside	Butte Valley	19	26	8	18	yes	3500	220	North Fork Feather River	
74	Swiss	" "	18	26	8	12		4500	50	" " " "	
75	Taber	Laporte	32	22	8	177		5300	160	Middle Fork Feather River	
76	Topaz-Centennial	Spanish Ranch	35	25	9	62		4800	204	Spanish Creek	
77	Union	Laporte	4	21	9	231	yes	5400	160	South Fork Feather River	
78	Union Channel	"	33	22	9	183	"	5400	160	" " " "	
79	United Placer([a])	"	6	21	9	242		5100		Slate Creek	
80	Volante	"	2	21	9	235		5500	67	" "	
81	Washington	Eclipse P. O.	31	23	10	154		6300		" "	
82	West Branch	Laporte	4	21	9	232		5400	129	South Fork Feather River	
83	Yankee Hill	"	10	21	9	240		5100		Slate Creek	
84	Young America	"	10	21	9	241		5100		" "	

[a] See Excelsior mine.

Register of Mines and Minerals of California—Placer Mines.

Plumas County.

	Name of Mine.	Nearest Town.	Location: Section.	Township.	Range.	Map No.	Whether Patented.	Elevation.	Area.	Length of Ditches.	Manner of Working.	Owner: Name.	Residence.
								Feet.	Acres.	Miles.			
1	Albion	Meadow Valley	32	24	8	98		3500	106		Ground sluic'g, etc	J. A. Edman	Meadow Valley.
2	Alturas (a)	Laporte	22	21	9	245		4500		2	Elevating & sluicing	Hendel & Packer	Laporte.
3	Amador (d)	Johnsville	18	22	12	219	yes	4800	320		Bank washing	Sierra Buttes Mining Co.	London, Eng.
4	Arrastraville	"	13	22	11	206		4850			Arrastra sites	S. Soracco et al.	Johnsville.
5	Bear Creek	"	11	22	11	203		5000	80		Ground sluicing	Mrs. H. Dunn	Johnsville.
6	Beatty	Spanish Ranch		25	6	44		2150			Sluicing in bed of c'k	J. Beatty	Spanish Ranch.
7	Belafer	Johnsville	7	22	12	215		4600	90		Sluicing on bar in c'k	Antone Belafer	Johnsville.
8	Brown Hill Extension	Cascade	31	22	8	175		5200	80		Ground sluicing	W. W. McMillan	Cascade.
9	Butterfield	Quincy	25	24	10	120		3800			" "	F. M. Butterfield	Quincy.
10	Calliope	Butte Valley	9	26	8	8		3400			Hoisting and sluicing	H. L. White & Son	Butte Valley.
11	China Ranch	Box P. O.	34	23	7	140		5600			Sluicing	Chinese	Box P. O.
12	Columbo	Johnsville	14	22	11	204		6000	90		Sluicing glacial gvls	Gimbroni & Co.	Johnsville.
13	Columbia	Quincy	28, 33	25	9	67		3900	150	2	Ground sluicing	R. Thompson et al.	Spanish Ranch.
14	Cook	Mohawk	24	22	12	220		4600	60		Sluicing	L. Christie	Mohawk,
15	Dutch Bar	Quincy	18	25	9	64		2800			"	Lucas & Co.	Quincy.
16	English Bar	Nelson Point	11	23	10	149		4050	80		Sluicing on river bar	S. A. Stewart & Co.	Nelson Point.
17	Everett	Quincy	7	23	10	146		4800				J. E. Wilson	Quincy.
18	Florian	Spanish Ranch		25	7	48		2250			Wingdam in riv'r bed	F. Gausner	Quincy.
19	French Bar	" "		25	7	49		2150			Sluicing on river bar	Ah Ping	Silver Creek.
20	Garibaldi	Johnsville	3	22	11	198		5500	120		Sluic'g on s'rface gvl	P. Lorenzi	Johnsville.
21	Goldbug	Cromberg	24	23	11	161		4200	40		Sluicing on river bar	Honck & Co.	Cromberg.
22	Hartery (b)	"	33	23	11	165		6000	50	½	Ground sluicing	Thomas Hartery	Quincy.
23	Hellas	Meadow Valley	33	24	8	99		3500	75		Sluicing creek grav'ls	J. A. Edman & Co.	Meadow Valley.
24	High Point	Nelson Point	7	23	11	156		4100	60		" shallow gvls	Stewart & Cook	Quincy.
25	Hillside	Cromberg	13	23	11	160		4300	40		" surface g'v'ls	John Benson	Cromberg.
26	Howard Hill	Johnsville	10	22	11	202		5600	160		Shallow " d'g'n's	J. R. Huff & Co.	Johnsville.
27	Jackson Creek	Cromberg	19	23	12	168		4500	160		Sluicing creek grav'ls	Jones & Sutton	Cromberg.
28	Jamison Creek	Johnsville	25	22	11	212	yes	5600	104		Glacial gravels	Jamison Mining Co.	San Francisco.
29	Jones, B. L.	"	6	22	12	214		4600	160		Sluicing creek grav'ls	Shinn & Jones	Johnsville.
30	Luman	Quincy	35	25	9	74		3500			Hoisting and sluicing	W. Luman & Co.	Quincy.
31	Luman & Dere	"	10	24	10	118		4000			Shallow gulch dig'ns	Luman, Dere et al.	Quincy.
32	Mill Creek	Spanish Ranch		25	7	53		2600			Ground sluicing	T. Johnson	Spanish Ranch.
33	Minerva Bar (c)	Nelson Point	18	23	10	151		3800			Hoisting and sluicing	Minerva Bar Mining Co.	
34	Morning Star	Johnsville	2	22	11	199		5000	160		Sluicing glacial gvls	A. Simonetti	Johnsville.
35	Mountain House	Spanish Ranch	31	25	8	59		5200			" surface "	Quincy Mining and Water Co.	Spanish Ranch.
36	Mountain View	Cromberg	29	23	12	173		5000	120		Surface mining	Engalls et al.	Clairville.
37	Nelson Creek	Nelson Point	22	23	10	152		4000			Sluicing	E. C. Fish	Oroville, Butte Co.
38	Outlet	Johnsville	1	22	11	200		4500	160		Sluicing creek grav'ls	J. M. Jackson	Mohawk.
39	Pine Leaf	Spanish Ranch	11	24	8	91		3900			Sluicing gulch g'v'ls	J. Shafer	Spanish Ranch.
40	Poplar	Cromberg	15	23	11	159		4300	120		" creek grav'ls	J. E. Lanfear	Quincy.
41	Ratford	Butte Valley	16	26	8	31		3100			Work on river bar	H. N. Ratford	Butte Valley.
42	Reising	Greenville	9	25	8	54		3100			Ground sluicing	J. Reising	Crescent Mills.
43	Santa Cruz	Quincy	19	24	11	123		4500			Sluicing surface gvls	D. M. Butterfield et al.	Quincy.
44	Shinn	"	26	24	10	119		3800			Washing creek g'v'ls	Shinn Bros.	Quincy.
45	Shoo Fly	Crescent Mills	3	25	9	63		3100			Work in river bench	R. Martin	Crescent Mills.
46	Silver Star	Spanish Ranch	15	24	8	92		3700			Elevating & sluicing	S. S. Taylor & Co.	Spanish Ranch.
47	Spring Garden	Quincy	30	24	11	128	yes	3900			Sluicing bench g'v'ls		
48	Verna	Cromberg	24	23	11	162		4200	60		Sluicing in riv'r bank	Wm. D. Vinny	Cromberg.
49	Washington (d)	Johnsville	24	22	11	208	yes	5250	140		Surface gravels	Sierra Buttes Mining Co.	London, Eng.
50	Weaver	Buck P. O.	12	23	7	136		5200			Ground sluicing	J. B. Jones	Buck P. O.
51	Woods Ravine	Spanish Ranch	7	24	9	108		3800	40	½	Sluicing tailings	R. Gifford	Spanish Ranch.
52	Yet Lee (e)	" "		25	7	254		2200	40		Derrick hoisting & sluicg	Yet Lee Mining Co.	Spanish Ranch.

[a] On creek 3½ miles. [b] Water supply 100 inches. [c] H. N. Egbert, agent, Nelson Point. [d] W. F. Johns, Superintendent, Johnsville. [e] 14 men.

REGISTER OF MINES AND MINERALS

PLUMAS COUNTY.

	PLUMAS COUNTY.		LOCATION.										WATER.							
	Name of Mine.	Nearest Town.	Section.	Township.	Range.	Map No.	Whether Patented.	Area.	Elevation.	Height of Bank.	Thickness of Gravel.	Source of Water.	Quantity, in Inches.	Width of Ditch.	Depth of Ditch.	Length of Ditch.	Grade of Ditch.	Length of Pipe.	Diameter of Pipe.	Head.
								Acres.	Feet.	Feet.	Feet.			Feet.	Feet.	Miles.	Inches.	Feet.	Inches.	Feet.
1	Badger Hill (a)	Spanish Ranch	7	24	9	109			4000	175	50	Wood's Ravine								
2	Bainbridge	Eclipse P. O.	30	23	10	153			6000											
3	Bean Hill	Spanish Ranch	3	24	8	85			4700	150		Silver Lake & Spanish C'k				10				
4	Bennet's (Brown Bear)	Wash P. O.	1	21	12	225		160	5300			Frazier & Mohawk C'ks				2				
5	Bernard's Diggings	Laporte	20	21	9	247			5000											
6	Black Rock (b)	Merrimac	3	22	6	184	yes		4900	60		Marble Creek								
7	Blue Gravel (b)	Nelson Point	10	23	10	250			4800			Willow Creek								
8	Blue Nose	Eclipse P. O.	12	22	10	191			5500			Nelson Creek								
9	Cascade	Genesee	14	24	11	122		160	5900	120	120	Clear Creek	800	3½				2000	11	250
10	Castle Rock	Nelson Point	10	23	10	150			4800											
11	Cement (b)	Mohawk	2	23	11	157		120	4400			Long Valley Creek								
12	Cleveland	Johnsville	3	22	11	197		120	6000			Squirrel Creek								
13	Concordia	Cromberg	18	23	12	167		400	5000			Long Valley & Jackson C'ks	2100	5, 7	3	10		1700	22, 11	300
14	Conly & Gowell	Laporte	16	21	9	243	yes		4900			S. Fork Feather River								
15	Consignee	Cromberg	30	23	12	172			4250											
16	Duquesne	Johnsville	4	22	11	196		80	6000											
17	Fairstake	Quincy	3	24	9	103			3800			Quincy M. & W. Co's D'h								
18	Forest Grove	Johnsville	9	22	11	201		60	6000			Nelson Creek				3				
19	Gard & Orr	Laporte	8	21	9	238	yes		5000											
20	Good Hope	Mohawk	7	24	12	216			4800			Jamison Creek								
21	Gopher Hill (c)	Spanish Ranch	12	22	8	90			3800	100	50	Silver Lake & Spanish C'k	2000			24		5000	22, 11	350
22	Grass Valley	Laporte	32	22	9	179		120	5000			S. Fork Feather River								
23	Grizzly Bear	Johnsville	3	22	11	249		168	5800	25	25	Bear Creek	300	4		1½		800	18, 12	250
24	Holloway	Gibsonville, Sierra Co.	11	22	10	190			5000			Hopkins Creek								
25	Hungarian Hill (d)	Quincy	17	24	9	111		500	4700			Slate & Gausner C'ks	1200							
26	Indian Hill	Spanish Ranch		25	7	51			2500	50		Kellogg Ravine	150			1		500	11	
27	Italian	Nelson Point	35	23	10	155			4500			Nelson Creek								
28	Jamison	Johnsville	31	23	12	174		120	4500	25	25	Squirrel Creek	600	4	2	¼	1	290	7	65
29	Kanaka Flat	Spanish Ranch	32	25	8	252			4800			Quincy M. & W. Co's D'h								
30	Kingsbury's Ferry	" "	22	25	8	57			2700	40	40	Kingsbury Creek								
31	Laporte Consolidated	Laporte	9	21	9	251			5100			S. Fork Feather River								
32	Lott	Chaparral, Butte Co.		25	6	42			5000			Gott's Lake, Chops C'k								
33	Michigan Hill	Quincy	26	25	9	68			3325	30	25	Cashman Creek	200	3		3½		600	15, 11	80
34	Miocene	Greenville	35	26	8	20			4600			Rush Creek								
35	Mount Hope	Chaparral, Butte Co.		25	6	43			5000			Ships C'k, Morris L'ke								
36	Northern Placer	Cromberg	6	23	12	166			5570	125		Little Long Valley C'k	1000	3	2½	3	⅝			210
37	Ohio Point	Spanish Ranch	18	25	8	56	yes	2700				Rush Creek								
38	Plumas Imperial	Quincy	19, 18	24	9	112, 113	"	1200	3800	100	100	Rock, Beaver and Deer Cks	2800			12¾		5000	22	
39	Red Hill (a)	Spanish Ranch	33	25	8	60			4700			Spanish Creek								
40	Richmond Hill	Eclipse P. O.	6	22	10	265		2400	5500			Onion Valley Reserv'r	2000			10				300
41	Ross	Butte Valley	16	26	8	19			3300											
42	Scad Point (b)	Meadow Valley	28	24	8	94			3550			Clear Creek	400	3		3		1000	15, 11	300
43	Secret Diggings	Laporte	21	21	9	246			4600			Rabbit Creek								
44	Smith's Hill	Spanish Ranch		25	7	50			2450			Oak Ravine								
45	Soracco	Johnsville	7	22	12	217			6000			Jamison Creek								
46	Spanish Flat	Laporte	16	21	9	244	yes		4800			S. Fork Feather River								
47	Star of Plumas	Buck P. O.	8	23	8	141			5000			Bear and Big Creeks								
48	Taylor Diggings	Genesee	21	26	11	36		200	5500			Davis C'k & Taylor Lake		3	3	12				150
49	Toland & Langdon	Spanish Ranch	9, 10	24	8	253			3800			Silver Creek				3			7	
50	Union Hill	Eclipse P. O.	3	22	10	186			6500											
51	Valentine	Spanish Ranch	5	24	8	84			4500			Quincy M. & W. Co's D'h								
52	Wilhelm Tell	Johnsville	7	22	12	218		80	6000			Jamison Creek								

[a] When hydraulicking takes water from Quincy Mining and Water Co.'s ditch. [b] Idle. [c] Worked by Wm. Kauradi under lease. [d] Drifting occasionally. [a] Quincy Mining and Water

OF CALIFORNIA—HYDRAULIC MINES.

Giants.	Drains into tributary of.	Men employed.	Date of permit granted by Debris Commis'n.	Owner. Name.	Owner. Residence.	Superintendent. Name.	Superintendent. Residence.	
1	North Fork Feather River	2	January 13, 1896	E. B. Jacks	Meadow Valley			1
	Middle Fork Feather River			Bainbridge P. & Q. Mining Co.				2
2	North Fork Feather River	8	1896	Quincy Mining and Water Co.	Spanish Ranch	A. B. White	Spanish Ranch	3
	Middle Fork Feather River			Andrew Jackson	Wash P. O.			4
	Yuba River	12				Ah Tie	Laporte	5
	Middle Fork Feather River							6
	" " " "		December 30, 1895	Plumas Blue Gravel Mining Co.	Nelson Point	Wm. Metcalf	Nelson Point	7
	" " " "			Nat Mullen	Eclipse P. O.			8
2	North Fork Feather River	8	November 6, 1895	Cascade Mining Co.	Vallejo, Solano Co.			9
	Middle Fork Feather River							10
	" " " "			J. Sutton & Co.	Mohawk			11
	" " " "			L. Sobrero & Co.	Johnsville			12
1	" " " "		October 14, 1895	L. V. Tefft & Co.	Cromberg			13
	Yuba River			Cox & Donahue	Laporte			14
	Middle Fork Feather River			Houck & Co.	Cromberg			15
	" " " "			R. R. Gumbert	Johnsville			16
	North Fork Feather River			F. C. Kaulback	Quincy			17
	Middle Fork Feather River	3		Curtis & Co.	Johnsville			18
	Yuba River							19
	Middle Fork Feather River			Ray Dolley	Beckwith			20
5	North Fork Feather River	4		Quincy Mining & Water Co.	Spanish Ranch	A. B. White	Spanish Ranch	21
	Middle Fork Feather River			C. S. Quigley	Laporte	C. S. Quigley	Laporte	22
2	" " " "	6	November 6, 1895	N. Rossi & Co.	Johnsville			23
	" " " "			Gould & Spencer	Gibsonville			24
	North Fork Feather River	2		J. H. Whitlock & Co.	Quincy			25
1	" " " "	3		Eyraud & Lazier	Spanish Ranch			26
	Middle Fork Feather River							27
1	" " " "		November 6, 1894	Northern Placer Mining Co.	Pittsburg, Penn.	P. Traynor	Cromberg	28
	North Fork Feather River	2	January 20, 1896	Wampler & Jacks	Spanish Ranch			29
	Middle Fork Feather River			O. D. Halstead	" "			30
	Yuba River		November 18, 1895	Laporte Consolidated Mining Co.	Duluth, Minn.	A. H. Holgate	Laporte or Oakland	31
	North Fork Feather River			C. F. Lott	Oroville, Butte Co.			32
1	" " " "	2	December 2, 1895	J. B. Higgins	Quincy			33
	" " " "			Whitney & Leck	Crescent Mills			34
	" " " "			—— Morris	Oroville, Butte Co.			35
1	Middle Fork Feather River		November 6, 1894	Northern Placer Mining Co.	Pittsburg, Penn.	P. Traynor	Cromberg	36
	North Fork Feather River			N. E. Duncan	Oroville, Butte Co.	J. Patten	Spanish Ranch	37
2	" " " "	15	July 8, 1895	Plumas Imperial Mining Co.	Minneapolis, Minn.	C. D. Hazzard	Quincy	38
	" " " "			Quincy Mining and Water Co.	Spanish Ranch	A. B. White	Spanish Ranch	39
2	Yuba River	20	November 18, 1895			A. H. Schnabel	Gibsonville	40
	North Fork Feather River			E. E. Ross	Butte Valley			41
1	" " " "	2		A. Robinson	Meadow Valley			42
	Yuba River	15				Ah Tie	Laporte	43
	North Fork Feather River			Cooley Brothers	Spanish Ranch			44
	Middle Fork Feather River	2		Steve Soracco	Johnsville			45
	Yuba River	12				Ah Tie	Laporte	46
	North Fork Feather River			Star of Plumas Mining Co.	San Francisco	F. Rood	Buck P. O.	47
2	" " " "	11		California G. M. & Investment Co.	Chicago, Ill.	R. C. Noyes	Genesee	48
	" " " "	2		Toland & Langdon	Spanish Ranch			49
	Middle Fork Feather River							50
	North Fork Feather River			Val. Roberts	Spanish Ranch	Valentine Roberts	Spanish Ranch	51
	Middle Fork Feather River	2		F. F. Vanzini	Johnsville			52

Company's ditch. Idle. [b] Now ground sluicing and cleaning bedrock.

APPENDIX III.

MINING REPORTS.

Contained herein are supplementary published accounts of Nelson Point mining activities:

On Nelson creek the stream has continued high, and the miners are but at getting in wing-dams and commencing their season's work. At the head of Nelson's some gold has been taken out. The Australia company are rich surface diggings, and have been taking out for a month past some $? a day to the hand. The English company are also doing well, making some $7 or $8 a day to the hand. A French company are taking out considerable gold, but as they keep their affairs to themselves, the amount of their earnings is not definitely known.- [Messenger.

(Mining and Scientific Press 5, Aug. 14, 1862)

MINING IN PLUMAS COUNTY.—The *Mountain Messenger* contains a long account of the mining operations in Plumas county, from which we select a few items as a speciman of what the miners have been doing. At Rich Bar $230 was taken out in a single lump. At Nelson Creek, 61 ounces were taken out by four hands in one day, and 37 ounces by four others...

(Hittell's Scraps, p. 26)

Nelson Creek.- Messr. Madden & Co. unearthed near $1000, with three hands, in one day last week.

Messr. Bryan & Co., with four hands, realized $700 in the first four days after commencing their claim...

(*Alta*, Jly. 7, 1854)

QUARTZ LEAD NEAR NELSON.—On Sunday last, Doc. Vaughn and Billy Frink took up a quartz ledge on a hill between the Illinois Ranch and Nelson Point, which promises well. It seems to be an extension of the Conkey ledge, struck several years since, and if so must prove valuable. From other prospectors we learn that the entire range of hills from Nelson Point to Rich Bar, on the middle Fork of the Feather River, are a succession of quartz ledges. Miners who worked there in '51-2 now remember ledges they blasted through in order to get to bedrock. The men are generally again hunting up their old stamping grounds and prospecting it as brisk and active as it was in the halcyon days when the white men first made their acquaintance of those regions.—*Plumas Standard*.

(*Bancroft Scraps* 51: 1-4, p. 224)

QUARTZ LEDGE NEAR NELSON.—Last Sunday, two men named Vaughn and Frink, took up a quartz ledge on the hill between the Illinois ranch and Nelson Point,

Plumas county, which promises well. The Standard says:- "It seems to be an extension of the Conkey ledge, struck several years since, and if so must prove valuable. From other prospectors we learn that the entire range of hills from Nelson Point to Rich Bar, on the Middle Fork of Feather River, are a succession of quartz ledges. Miners who worked there in '51-2, now remember ledges they blasted through in order to get to bed rock. These men are, generall, again hunting up their old stamping grounds and prospecting is as brisk and active as it was in the halcyon days when the white man first made the acquaintance of those regions"

(*Butte Record*, Nov. 15, 1862)

MINING IN PLUMAS.—A correspondent of the *Plumas Standard*, writing from Nelson Point, August 31st, says:

The miners here are doing well as could be expected in old diggings that have been worked five or six times over and then handed over to the Chinese and worked I don't know how many times. Thorpe & Co. are averaging five dollars a day to the hand. Johnson, Bushman & Co.'s claims at Independence Bar are paying good wages. Harmon, Oliver & Co. at Henpeck City are making seven dollars a day to the hand. Jesse Brown & Co., alias the Peanut Company, are making small wages, but have not as yet got to the bed rock. Uncle Samuel Blair & Co., Hopkins creek, three hands, took out between four and five hundred dollars last week. These are all claims in the bed of the stream.

(*Bancroft Scraps* 51: 1-4, p. 224)

The following list proves that there is no "let up" to old Nelson, and that the mining claims in that section are "good for wages" for some time yet. Bushman & Co.'s claim is paying regularly $20 per day to the hand. They have five men at work. Carr & Co.'s claim averages $15 per day to the hand; Gildersleeve's, $7. McEwer & Co. have just reached the bed rock in their claim, and some of the dirt near the rock will pan out one ounce to the pan. Nearly all of the other claims are paying good wages...

(*Quincy Union*, Sept. 18, 1864)

APPENDIX IV.

CORRESPONDENCE.

Contained herein is correspondence from Nelson Point that was published in the *Plumas Standard* and the *Quincy Union* between 1860 and 1866.

Correspondence.

Old Nelson Took In.

Nelson Point, January, 1860.

Editor Standard:—I have been waiting for Micaw. ber like, for something to turn up, to write about for more than two weeks. The severe cold weather, interspersed as it has been with an occasional snow storm, has made the times here as Mark Tapley would say "very jolly indeed." The little town is as quiet as a Quaker's meeting, and as dull and gloomy as a funeral. Newspapers are above par. Let me tell you the mountain newspapers have to pass through a terrible ordeal these times. A person will catch up a newspaper, read it all over carefully and throw it down. A few moments after he will take the same paper again and commence to read just as if he supposed the paper to constantly manufacturing more news. Reading over and over again will make even the ablest article become dull and insipid. The *Standard* endures this test very well. I have frequently heard it spoken of favorably and never disparingly. But hold on; it was not my intention to pen a puff for the *Standard*, and I will stop where I am.

I attended our Theatre last evening. The performances were so novel, the play entirely new, and as I have the best of reasons for believing will never again be produced in the State, I have determined to give you a description. I hope I may have a full, free pardon extended to me for being prosy. The title of the play was "The Grand Mountain Steamboat—or the way to do it;" and allow me to say it was done to perfection. The audience was not large, but very select indeed, being composed entirely of the *elite* of the *beau monde*. The scenery we never saw equalled in any theatre. The music I am sorry to say was an entire failure in consequence of our celebrated violinist, the man that plays the tune of the stump tailed dog so beautifully not being at his post. The price of admission was "nary red," and perfectly satisfactory. The curtain rose at early candle light and exposed to our view the beautiful scenery of a store well stocked, and an actor who played the part of clerk in said store. The actor of the evening Stranger, enters and bows very politely. He is grace and ease itself, well dressed but not foppish. He compliments the clerk and the store in an easy confidential way quite charming. Presently, all by accident, he remarks "now you don't happen to have any choice liquors on tap." The modest clerk does though—stranger prefers brandy, drinks and remarks there is no such brandy as that in Quincy, and down goes the curtain. The curtain rises again and shows a mountain hotel as natural as life, and an *hombre* officiating as landlord, cook, &c..

Stranger enters the hotel with a polite bow and as much grace and ease as my Lord Noodle, and orders supper. Landlord and cook is flustered, but flies to the kitchen...Bless me how the choice delicacies of the mountains accumulate on the table. Stranger eats and compliments everything. Remarks no such meal as that can be had in Quincy. Cook and landlord is pleased with the compliment—Stranger regales himself with a cigar in the bar room that he says beats Quincy all hallow. The curtain now falls and when it rises again reveals a billiard saloon as natural as life. Even Hezekiah's chicken, in a cage, is dimly visible in the back ground. The characters here are saloon man, barkeep, and some half a dozen patron saints. There they sit toasting themselves by the side of a red hot stove with heels elevated at an angle of forty-five degrees, puffing away at a cigar or a stale pipe, dozing over a newspaper, or fast asleep. Presently one will gape, then another and another, till all that are sufficiently awake have gaped. Then one very sagely remarks that its extremely dull, and a few minutes another remarks how very dull, and so that goes the rounds. The graceful stranger enters, bows, and scrapes as no one else can do and all is life and animation where it was so dull and gloomy only a moment before. They are disconcerted a little at first, but stranger soon makes all at their ease around him. Barkeep is fully impressed with the importance of stranger, and stifs the fire desperately, and touches up the camphene lamp. Never, never did that lamp throw out such a brilliant dazzling light before. A bummer asks barkeep for tobacco,—gets a chew, and then mischievously shows barkeep a whole plug. Stranger immediately asks bummer for tobacco—takes a very delicate chew and soon desires another. Bummer in the fullness of his generous soul tells stranger to take a piece. Stranger with great dexterita cleaves the plug of tobacco at the very center and pockets half, much to the disco????? of poor bummer. But it is better tobacco than can be bought in Quincy, Stranger says, and that is some consolation. Saloon man treats man again. Stranger compliments everything this time in the most eloquence. The brandy is really delicious, the billiard table superb, and the saloon a perfect palace. There is nothing in Quincy to be compared with them. Stranger banters the saloon man to play billiards, and at it they go game after game. It is now stranger and now saloon man. Barkeep fires up, and touches up the aforesaid lamp with a skillful hand. A fall piece of humanity stands hour after hour waiting to escort the stranger to his room at the hotel, and has become the very picture of despair hours ago, but still on, on the game goes, regardless of the waning night and fast approaching day. The game has run into the thousands—stranger has been treated times enough and finds himself sufficiently amused. He bows very low to the saloon man says good evening, sir, and adds, in charming accents, "I will see you, sir, in the morning," and walks very gracefully out. Saloon man is stupified for a moment. Barkeep says, "follow him." Saloon man takes three steps toward the door very quick—hesitates—stops and returns to his place beside the stove, awfully crestfallen. Stranger is wrapt in the arms of old Morpheus in the best room at the hotel, when the curtain to our great stage falls and closes the scene, I neglected to mention of, but stranger modestly hinted in the early part of the evening a preference for an early start in the morning. The curtain rises again, and never did mortal eyes behold a better representation of morning. It reveals landlord and cook in a flury, but with his cloth slippers on moving as quietly as a mouse, lest he shall disturb the gentle slumbers of stranger. Stranger rises, partakes of his sumptuous repast and showers compliments. There is no such breakfast to be had in Quincy, oh no! Stranger says this to himself. Finally he takes a very affectionate leave, saying very sweetly, "I will settle with you, sir, when I return." Landlord and cook answers very sharply, I don't keep books, and gives stranger a frosty look that would have instantly killed a whole patch of pumpkin vines, but stranger makes his exit not in the least disconcerted, and humming a pretty little tune as he goes. As he rises the mountain in the back ground he approaches a man sawing wood, whom he banters for a pair of snow shoes, pay when he returns. The man does not wish to sell, and when stranger moves on, he places his hand to his face as you have seen mischievious boys do before now, and stranger disappears.—Never was mortal man more popular than the star actor of last evening at old Nelson. Stranger is indeed eminently qualified to travel, and has sufficient brass in his face for a lower country politician. We like him very much. He teaches a very excellent lesson. I do wonder if men will ever learn that a tasty garb and genteel bearing has in reality so more to do with a man's honor and honesty than chalk has to do with cheese. We have suddenly become a very genteel people, here, if polite bows are a sure mark of gentility. But its time to close this.

Yours, &c., HEZEKIAH BUMMER.

LETTER FROM NELSON POINT.

NELSON POINT, June 12, 1865.

ED. UNION: Quincy with her charms has "gone from my gaze," and I again find myself on Feather River, enjoying camp-life very much as of old, if not less, and reveling in high hopes of making a strike, which is also the earnest prayer of my creditors. I have lately been engaged in prospecting and came to the conclusion that that particular branch of mining only pays when rewarded in the end by rich diggings, or somebody has better faith than the prospector, and relieves him of all anxiety about the future development of his claim, by buying him out. Neither of the two conditions being likely to be carried into effect in my case, I "took up my bed and walked,—pre-empted a quarter section of Feather river (not yet surveyed, but in market) and propose, without a chance of failure either to make a stake or burst. Without apprehension of the latter, I will now proceed to jot down the few items of probable interest of this vicinity.

Where river mining is the occupation of nearly three-fourths of the mining population no accounts of large yields are recorded until the latter part of the Summer or Fall, therefore the pleasure of regaling you on thousands of dollars made by Tom, Dick and Harry, you must be deferred until some time after the "Fourth." The Middle Fork of Feather River for about three miles below Nelson Point is to be thoroughly cleaned out this summer. No less than five flumes, besides numerous wing-dams are in course of construction between the mouth of Nelson Creek and Shasta Bar. The banks of the river present a busy appearance. The miner for the season becomes mechanic; pick and shovel are laid aside, and then—well, he resumes pick and shovel again and expects every day to get a small fortune to the pan, which expectation is never realized, but relieves a great deal of fatigue, caused by tenaciously holding on to a buggy which invariably upsets at the end of the road it is traveling on. Two of the claims now being flumed, one called the Fish-trap; the other owned by Maguire & Co., were worked profitably last year, and prospects obtained before the fall freshets suspended operations leave scarecly a doubt of their paying equally as well this summer. Of the others there are none but what are expected to yield every share-holder a handsome fortune, computed to suit each man's ideas of what constitutes that desirable sum.

Mitchel, Giles & Co., are engaged in running a tunnel in a hill adjacent French Ravine. They are within a few feet of the shaft, from which they obtained their prospect, three bits to the pan. Copeland & Co., struck good prospects, one dollar to the pan, in a cut sluiced in the hill opposite Nelson Point, unfortunately for them the water failed before the claim could be farther developed. Bonner, Hardy & Co., have been piping all winter in the old Lewis Ranch. They intend to clean up shortly. Their ground-sluice is estimated to contain about six thousand dollars.

Nelson Creek is rapidly passing into the hands of Celestials. I have heard of but four or five companies of white men intending to work the Creek the coming summer.

The Ancient and Honorable order of the E.V.C., held a special meeting last night.—Business of importance was transacted and new officers were elected. Since the opening of the road over the mountain, travel has assumed a healthier aspect, and daily new faces present themselves at headquarters. Whether they happen here by chance, or are attracted by the well deserved fame of Old Nelson for rich diggings, I do not profess to know. The boys finished a course of lessons in dancing, a few nights ago, and I suppose are now ready to attend all the parties in the neighborhood, provided they are given before they get out of practice.

T.

LETTER FROM NELSON POINT.

NELSON POINT, Oct. 17, 1865.

EDITOR UNION:—The material for a communication is unusually scarce. With but few exceptions, the mining news is not encouraging, as far as regards this immediate neighborhood. But as Pat says: "Bad luck is better than no luck at all." I will give you a short history of this Summer's operations, although the failures outnumber the successes:

Follow me, if you please; I grant it is quite a favor, for nobody would walk the road for pleasure; but this is for information, and that road is more or less uneven.

The first claim you see flumed, is the old Island claim. It was flumed three times before,—twice to the discomfiture of the parties engaged, and promising no better this year. The next, owned by Hill & Co., at the lower end of Columbia Flat, is a *quien sabe* case,—won't pay assessments, I am informed.

We are now at Rocky Bar. Of the two Chinese companies engaged here, the upper one has made good wages, and the lower one—nix.

Next in order is the claim of Jack o'Diamonds. I am credibly informed this claim has almost paid grub and whiskey.

The Fish-Trap is the next. It is one of the exceptions,—has pretty nearly paid expenses, and promises a dividend to shareholders.

That long flume right below Missouri Falls is owned by two companies, who joined flumes. All the difference in the claims belonging to the companies, respectively, is, the lower one paid, and the upper one did'nt.

Now comes in quick succession a number of wing-dams, all belonging to Celestials, yielding—I guess they know—but I don't.

We will go to the next claim, which is owned by the Double-geared Lightning Company. It is a remarkable fact, that this company have, with their railroad and hoisting apparatus, not only moved more dirt and worked more ground than any company on the river, but have also made less money than any other.

Now, Dear Union, I beg to be excused from going any further down the river. We have not found anything very good this far, and I'll assure you that the four or five Chinese companies below Rich Bar and Baker's Point, have no better claims than those above mentioned.

Returning to Nelson Point, which retains its ancient aspect, (I always know it at sight, even after an absence of weeks,) we will turn our attention to Nelson Creek:

Marken & Co., abreast of Fiddler Flat, worked a water-derrick all Summer, hoisted all their dirt and washed it, paying average wages; and Chris. Berk has done a good Summer's work in the old Saw-mill claim. These two, I believe, are the only claims, owned by white men on the creek, that have compensated their owners for the labor expended.

French Hill stock is coming up, since Cooksey & Co. struck paying gravel in their tunnel. Other companies are organizing with a view of tapping the hill adjoining the claims of C. & Co.

I would close this letter with a few news items, outside of mining, but unluckily for paragraphs, everything goes on the usual way. Ours is a peaceable and law-abiding community, and consequently I have no tales of horror or roguery to unfold. To draw an imagination, lays me liable to the charge of "adding." Moderation in all things, is a good motto, and I think you have had a pretty copious draught of T.

LETTER FROM NELSON POINT.

Nelson Point, Oct. 25, 1865.

DEAR UNION:—While pondering on the vicissitudes of life, it occurred to me that some of my worthy bethren of the E.C.V., while engaging in mining this Summer, were badly sold. I sympathize with them; but this sympathy does not only exist in the heart, but also in the pocket. Some of the river companies have suspended work this week. You can easily tell who they are, for they all sing

"Curse the luck, curse the luck," etc.

Those who are still at work, with a look that plainly shows their incredulityof getting their hopes converted into cash, still retain the original:

"Trust to luck, trust to luck-

Stare fate in the face," etc.

I am one of the latter, but have an idea I'll change the tune ere long, and adopt the new version.

It rained hard this morning, and I prayed that the flume might fall down; but some righteous individual must have prayed against me, for it quit raining, the flume still stands, and the work goes bravely on.

A recruiting agent of the Mexican Government, who must have heard of some of the boys going up the flume, eyed me rather sharply the other day, and thought, no doubt, "That *hombre* will do;" but I did not do it, and so lost my chance of attaining martial fame.

While earthquakes are going the rounds, I would advise a few to come up the river, (one won't do any good,—the gold is too scattering,) to save us, as the Era says about balloons and earthquakes, from "going up—up—up." Nothing else but an earthquake can resurrect the river diggings of this section, and it will be hard work for it. I, for my part, give them up for dead, and shall go prospecting for copper and zinc to make brass. Enough of that will help a man when he has no other chink.

Since mother earth has been in convulsions (and nobody has got well over the shocking figure she cut,) I have directed an occasional glance at the supposed-to-be extinct crater, immediately in my rear, or rather the rear of the house I am occupying, to detect some phenomena preceding an eruption, but all is quiet on the mountain yet. This crater is about three miles below Nelson Point. The base of the mountain forms what is known as Rich Bar. The crater, or opening, is in the shape of a half moon, at an elevation of about fifteen hundred or two thousand feet above the river. A man by the name of Beard descended to the depth of two hundred feet in '52, where he found a bench in the rock. He brought up round pebbles and stalactites, taken from the sides at that depth. In consequence of the close proximity of this dangerous (?) pet and other causes combined, I came very near having the blues the other day, in spite of hard work, which is said to be an effectual remedy for them. But while thinking whether to have them or not, I caught myself singing.

"Oh! let the world jog along as it will, I'll be free and easy still," etc. A few stances cured me of all symtpoms. If any of your readers should be likewise affected, give them the above without fear of being sued for mal-practice.

Since writing the above, I have heard a friend of mine offer to sell his Summer's work for five dollars, two bottles of whiskey and a pound of grapes. The offer was rejected, and in order to effect a sale myself, I am compelled to make a lower offer, namely: A young widow with three or four children, and a well improved ranch. For further information, apply to the undersigned. To make up for the river failures, the boys making preparations for a vigorous Spring raid on hill and bank claims. Nearly all of them prospected well at the time the water failed. The chances are they'll all be happy yet. Yours, &c.,

NOMEN NESCIO.

CORRESPONDENCE.

NELSON POINT, Nov. 26th, 1865.

DEAR UNION:—For the last few days it has been a doubtful case whether one would get up in the morning, or not; the wind has blown furiously, mostly of nights, felling trees in all directions and playing smash generally. Tall, stately pines, surrounding one's dwelling, may be very good for ornament or shade in Summer, but of nights, when old Boreas is in a bender, rather than be kept in constant fear of being reduced to mincemeat, I would prefer to have them transplanted, and dispense with ornament.

The late rains raised the river some ten or twelve feet above low water mark, making a clear sweep of all flumes and wing-dams remaining in the river. No sign is left whereby to tell where men have toiled and slaved; some gainers in money and experience in mining by it, others, only the latter. Obliterated are the marks that Unlucky & Co. pointed out as the scene of their late disaster. Nothing remains to them but sundry slips of paper, with the song "Promise to pay," engraved upon them, entrusted to them, with the general understanding that the trustee is not to keep them for an indefinite period of time. My advice to them, is the same as that given to the man who got fast in the cave: "If you pushed in, you can kick out,"

A Chinese wedding was celebrated at Sterling Flat yesterday. The ceremony was conducted in American style, and attended by a number of that *genus homo* from the Point, who thought it appropriate to kiss the bride. The last part of the ceremony must have been interesting indeed. Oh, ye gods!—that chapter on kisses would be appropriate. When can the heart be moved with more rapturous emotions? Oh, how sweet how tender must a kiss from Celestial lips!—ugh!

Any man, claiming to be a philanthropist, would second the building of a wagon road from American Valley to Sawpit Flat, via Nelson Point. The grunts escaping from mules under heavy burden are heart-rending, without the continuous stream of blessings poured on their devoted heads by those attending them. If ox-teams traveled the road, the latter pious expressions would become intelligible by being forcibly translated from jargon Spanish to Missourian. The luminous idea of facilitating travel and trade with a neighboring county is hard to get through a Board as long as Road Funds are low; but this is a bad time to talk about it, so we will wait till Spring before renewing the subject.

Mining has gained some impetus since the rain, nearly everbody being at work making the best of the water before it freezes up.

Many are the ways proposed for enjoying the long Winter evenings; debating, singing, and dancing schools are all in the programme.

Yours Truly,
NOMEN NESCIO.

CORRESPONDENCE.

BACHELOR'S HALL, NEAR NELSON POINT,
Dec. 5th, 1865.

A BACHELOR'S TRIALS.

EDITOR UNION:—For the past day or two I have seriously thought of getting married. If it was not for apeing that Chinaman who got Benedictized here the other day, I believe I would, provided-. That dash is easily interpreted by any old bach. of thirty odd, who has popped the question more than once with the same result;—another dash, which stands for a sigh of relief. Confound that cat!—I verily believe she jumped into the wash-tub. "Accidents will happen," but she has no business to practice gymnastic feats,—a natural consequence of living in a state of singleness. (I believe singleness is the word; it means simplicity; whether on my part, or the cat's, I don't know). "Here cook! Is dinner ready?" That's simply a hint that I am chief cook and bottle-washer this week. In the discharge of my duty as head of the culinary department, I forgot the cat, and left her in the suds. Oh, who would not be a bachelor! How delicios your grub tastes when you cook it yourself! What a blessing to have some to cook. Cold? Did you say it was cold? Well, why don't you button your shirt collar, bosom and sleeves? For the simple reason, my dear sir, that the shirt parted company with those valuable appendages, buttons, some two or three weeks ago; but I'll sew them on next Sunday, if I don't go to town. Rip! there you go again! tore my unmentionables from stem to stern, and not a needle-full of thread in the house. Going to town, Dick? Well, bring me a bit's worth of thread, a dollars worth of needles and a pair of pants. Pants arrive just big enough for two, of my size. I know a remedy for that—take a double reef in the waist-band and cut six inches off the bottom; that's what's meant by "curtailing expenses."

Last night was a bitter cold one; I slept cold, and two or three bunches of Irish feathers did not add any to my comfort, besides proving neglect in my private affairs.

The question recurs, whether it is cheaper to splice and quit rum and tobacco, or, by pursuing the old way,—rough it.

Until I get married, I'll continue to sign my maiden name,

NOMEN NESCIO.

LETTER FROM NELSON POINT.

NELSON POINT, Dec. 30, 1865.

DEAR UNION:—About the time this reaches you, eighteen hundred and sixty-six will be ushered in. Though delinquent in many other things, I'll not fail in wishing you a happy New Year. Another day, and 1865, with its memorable 3rd of April,—but, alas! also its 14th—will have passed away. Long will be remembered the momentous events that have transpired in its course. It dates the assassination of our noble President, but, thank God! also the dying throes of rebellion. Many brave boys yielded their lives in its beginning to make the end a happy one for us.—Then, while the cup is full of joy over the end achieved, let us drink in silence to the memory of the departed heroes.

Christmas—Merry Christmas—has gone, but is coming again, they say. The boys enjoyed it hugely, seemingly for no other reason but that it, like the Fourth of July, comes but once a year. Shooting for chickens, was one way of enjoying it, and eating them, another,—the latter by far the most agreeable; the weather not being what is termed very fine, hence, in-door amusements were preferable to out-door exercises. We had no ball on Christmas, and I am assured there will be none on New Years; a majority of this community are so strictly orthodox, that dancing is scarcely thought of, and condemned *in toto* by all but four or five ladies. It is customary to have religious service in the forenoon on Christmas day, but this flock was deprived of that spiritual comfort, on account of the minister's absence.

Owing to the holidays, my head is not quite clear on the subject of mining, or any other news items. The cold weather has reduced the water so much, that little mining is done, except in "drift diggings."—Foremost of them are those of See & Co., in the hill opposite Nelson Point. About two or three miles of this hill is open for claimants. The miners around here are very considerate, refraining from laying claim to it, while they have moderately good claims, in order to give strangers, wishing to locate here, an excellent chance to enrich themselves. Giles, Oliver & Co., struck better pay-gravel than ever in their tunnel in French Hill, and have commenced "breasting out." Jackson, Mitchel & Jolly discovered good diggings at the mouth of Haun Ravine; an average of twenty cents per pan was obtained in the gravel. A number of men are at work in Willow Creek, making wages. New mines have been found on Bachelor Creek. Fell & Co. have established themselves at that rock-bound retreat, with fair prospects of enjoying the utmost quiet in that secluded spot.

NOMEN NESCIO

LETTER FROM NELSON POINT.

NELSON POINT, Feb. 2nd, 1866.

EDITOR UNION:—The weather permits one the quiet enjoyment of all day in the house. The amount of enjoyment derived, generally dependes on the company you have. I can enjoy myself hugely by listening to the conversation of others, but particularly to the discourse of a lady when directed to love affairs and other kindred subjects. To be a good listener is a virtue sometimes, when patience is taxed to the utmost by somebody boring you to death with recitals of his lordly self. It requires attention to put in the monosyllables, yes and no, in the right place and time. But I have disgressed—I was going to tell you about the weather—did you ever hear any body discourse on the state of the weather? I suppose not, and in the attempt to pick up something new, I hit upon it. But after all the subject is not a long one; I pronounce it decidely mixed—snowing one minute, raining the next, then a slight decoction of hail and sleet, with an occasional beam of sunshine interwoven, are its prominent features,—which keep one's own features carefully hidden away under shelter; and while you see the countenances of the boys peeping out from their loop-holes, you will remark the evident change that has taken place in them. Elongated to twice their length before, fearful of a repetition of the two preceding Winters, they have now assumed their proper shape, with a twinkle of delight added in anticipation of a good time coming,

Our river was up to a considerable hight, but has subsided again. By saying *our* river, I don't mean that I lay any claim to it. Far from it; I had all I wanted of it last Summer. It lacked some ten or twelve feet of reaching the high-water mark of '61 and '62 No particular damage was done, as far as heard from-a natural consequence around here, where there is nothing to lose.

A petition was circulated freely this last week, praying the Legislature to authorize the issuing of bonds for the purpose of building a wagon road from Quincy via Nelson Point to Slate Creek, and so on out of the wilderness. The amount asked for ($15,000), is too-small, in my estimation, to accomplish the whole, but does well enough for a beginning.

Yours, Mud-bound,
NOMEN NESCIO.

LETTER FROM NELSON POINT.

NELSON POINT, Feb. 11th, 1866.

EDITOR UNION:—Your surmise of dull times all over the county, but not receiving any communications, was, perhaps, correct, and the cause thereof was, no doubt, the

RAINY WEATHER.

There is such a thing as getting surfeited of a good thing. Of course everybody engaged in minig anxiously wished for rain or snow to supply him with water, but to have the floodgates opened for the space of forty days and nights was no part of their calculation. But it must have rained a good deal harder at the time of the deluge, since the hill-tops are still out of the water. That's some consolation—Man proposes, God disposes—so if it rains any more we will have to do as people do in the Atlantic States—let it rain and be—contented. This continual drainage from above is very inconvenient to

TRAVELERS.

Pedestrians, in particular, have a hard time taking soundings in the mud. Even miners, who delight working in the dirt. (poviding it is rich), do not relish walking in it. I saw a Chinaman who was drifting to leeward in No. 15 mud-scows strike, stern first; he got up again and seemed to sail smoothly in deeper mud. His course was shaped for your humble correspondent's "country residence," which proved to be another stumbling-block, in fact,

A MISTAKE.

He lost his reckonings—sailed to the wrong port. He was in quest of chickens! But is it a wonder? Nobody but a fool or a Chinaman would have asked me if I had chickens to sell! From my appearance, he would not live a minute within reaching-distance of me, or, that me name wasn't John.

THE MINES.

From what I can learn, are very productive. You know I am trying to pick up news-items for the Union, and as a general thing I am pretty inquisitive, in order to be among the first to learn of new strikes. On meeting one of the boys, I asked him,—How do you do? and nine out of ten tell me, "Pretty well;" the tenth one bores me by complaining of illness, but, as that is no answer to my question, I think he is making money and trying to hide it by evading an answer. It is needless to enumerate the companies; they are "doing" all alike,—well, you know. One man told me he was doing as he pleased; I hardly know how that is, but I reckon not over an ounce a day.

CHIT-CHAT

I have not seen anything of your Nelson correspondent, Nomen Nescio, lately, and presume he is dead. This is only a presumption, but I know he is dead in more ways than one,—dead-head, dead broke, dead in-love, etc; but for fear he'll deaden me if he sees this, I'll close by signing myself, his most

Devoted Friend, T.

NELSON AND VICINITY.—A correspondent writing under date, Oct. 21, sends us the following:

Ellerman & Gohda were twice driven out of their claims by the high water, but are and have been doing well. Stevens, Ballou & Co. were also driven out. They have fitted up again, and have a number of Chinamen working on shares. Reed & Kane have also resumed operations and are making from six to ten dollars per day to the hand, and have a prospect of doing still better. Koher, Lee & Co., have commenced work again. They are down twenty feet, and find the "bed stun" still pitching, up stream.—Things look rather "scaly" for the boys, and they think of going to the dry diggings before the rainy season commences. The Chinese companies in the river are making money this season, at least dust seems to plenty with them, and they are living on the fat of the land—(hog meat and chicken.) Shubert & Murken are not mining at present, but are putting up a dwelling house. After they get fixed up, they intend to get out lumber, timbers, &c., for the bed rock flume in Nelson Creek. which promises to be one of the best paying investments ever made on the creek, or in the county. Jackson & Jolly have their claims nearly fitted up, and will be ready to commence work as soon as a supply of water can be had. Their claims paid good wages last season, and they hope to do still better the next.—Hardy's claims are paying him good wages; Dilks' the same. The "Duke" and Sam. Ballou have just completed a ditch to carry water on some new claims. , which the "Duke" says are the best in the country. Our little town has presented quite a lively appearance during the past three or four months, but as the road and bridge are completed, and the men about to leave, it will doubtless be dull again for a while. Social and *honest* games are occasionally indulged in by some of our citizens to pass off the evenings.—Even the native Americans once in a while indulge in the same pastime, but as a matter of course, always get `nipped.' A little scrimmage took place here a short time ago, between a couple of our notables. Each party got enough of the pi, and quit of their own accord. One had his figure-head disfigured, his sky-light darkened, and his cut-water bruised, while the other fared but little better. Teams have commenced to travel on the new road. We presume the the town authorities will after due consideration of the subject, conclude to remove the town to some point nearer the road.

APPENDIX V.

MINERS' TEN COMMANDMENTS.

THE MINERS' TEN COMMANDMENTS.

A man spake these words and said :—I am a miner, who wandered " from away down east," and came to sojourn in a strange land and "see the elephant." And behold I saw him, and bear witness, and that from the key of his trunk to the end of his tail, his whole body has passed before me ; and I followed him until his huge feet stood still before a clapboard shanty ; then, with his trunk extended, he pointed to a candle-card tacked upon a shingle, as though he would say read, and I read.

THE TEN COMMANDMENTS.

I. Thou shalt have no other claim than one.

II. Thou shalt not make thyself any false claim, nor any likeness to a mean man, by jumping one ; whatever thou findest on the top above, or on the rock beneath, or in a crevice underneath the rock—for I am a jealous dog, and will visit the miners round with my presence, to invite them on my side ; and when they decide against thee, thou shalt have to take thy pick, and thy pan, thy shovel and thy blankets, with all thou hast, and go "prospecting," both north and south, to seek good diggings ; and thou shalt find none. Then, when thou hast returned, in sorrow thou shalt find that thine old claim is worked out, and no pile made thee, to hide in the ground, or in an old boot beneath thy bunk, or in buckskin or bottle beneath thy cabin, but hast paid all that was in thy pures away, worn out thy boots and thy garments, so that there is nothing good about them but the pockets, and thy patience be liked unto thy garments ; and at last thou shalt hire. thy body out to make thy board and save thy bacon.

III.—Thou shalt not go prospecting before thy claim gives out. Thou shalt not take thy money, nor thy gold dust, nor thy good name, to the gaming table in vain ; for monte, twenty-one, roulette, faro, lansquenet and poker will prove to thee, that the more thou puttest down, the less thou shalt take up ; and when thou thinkest of thy wife and children, thou shalt not hold thyself guiltless but insane.

IV.—Thou shalt not remember what thy friends do at home on the Sabbath day, lest the rememberance may not compare favorably with what thou doest. Six days thou mayest dig or pick all that the body can stand under ; but the other day is Sunday, when thou shalt wash all thy dirty shirts, darn all thy stockings, tap all thy boots, mend all thy clothing, chop thy whole week's firewood, make and bake thy bread and boil thy pork and beans, that thou wait not when thou returnest from thy long tour, weary. For in six day's labor only thou canst not work enough to wear out thy body in two years: but if thou workest hard on Sunday also, thou canst do it in six months ; and thy son, and thy daughter, thy male friend and thy female friend, thy morals and thy conscience be none the better for it; but reproach thee, shouldst thou ever return with thy worn out body to thy mother's fireside, and thou strive to justify thyself, because the trader and the blacksmith, the carpenter and the merchant, the tailors, Jews, buccaneers defy God and civilization, by keeping not the Sabbath day, and wish not for a day of rest, such as memory, youth and home made hallowed.

V.—Think more of the gold and how thou canst make it fastest, than how thou wilt enjoy it, after thou hast ridden roughshod, over thy good old parents' precepts and examples that thou mayest have something to reproach and sting thee, when thou art left *alone* in the land where thy father's blessing and thy mother's love hath sent thee.

VI.—Thou shalt not kill thine own body by working in the rain, even though thou shalt make enough to buy physic and attendance with—neither shalt thou kill thy neighbor's body by shooting him, except he give thee offence— then upon the principle of honor, without principle, thou mayest, even though by "keeping cool" thou hadst saved his life and thy conscience.

VII.—Thou shalt not grow discouraged, and think of going home before thou hast made thy "pile" because thou hast not "struck alead," nor found a "rich crevice," nor sunk a hole upon a "pocket," lest in going home, thou shalt leave $4 a day, and go to work, ashamed, at 50 cents, and serve thee right ; for here, by staying, thou mightest strike a lead and fifty dollars a day, and thy manly self-respect, and then go home with enough to make thyself and others hapy.

VIII.—Thou shalt not pick out specimens from the company pan, and put them in thy mouth or in thy purse. Neither shalt thou take from thy cabin-mate his gold dust to add to thine, lest he find thee out, and straightway call his fellow-miners togeather, and they hang thee, or give thee fifty lashes and two hours to leave the country, or brand thee like a horse thief with R upon the cheek, to be "known and read of all men"—California in particular. And if thou steal a shovel, or a pick, or a pan, from thy toiling fellow-miner, hanging would be too good for thee, and thou asked to be kicked and cow-hided for thy pains ; and forever hang down thy head.

IX.—Thou shalt not tell any false tales about "gold-diggings in the mountains" to thy neighbor, that thou mayest benefit a friend who hath mules, and provisions, and tools, and blankets he cannot sell— lest in deceiving thy neighbor, when he returneth through the snow, with aught save his rifle, he present thee with the contents thereof, and like a dog, thou shalt fall down and die.

X.—Thou shalt not covet thy neighbor's gold nor his claim, nor undermine his bank in following a lead, nor move his stake, nor wash the tailings from his sluice's mouth, nor throw up dirt upon his bank. And if thy neighbor have his family here, and thou love and covet his daughter's hand in marriage, thou shalt lose no time in seeking her affection ; and when thou hast obtained it, thou shalt " pop the question" like a man, lest another more manly than thou art should step in before thee, and thou covet her in vain ; and in the anguish of disappointment, thou shalt quote the language of the great and say, " Let her rip !" and thy future life be that of a poor, lonely, despised, and comfortless bachelor. The end.

A LITTLE ONE THROWN IN.—Thou shalt not dig up a public road, unless thou canst afford to fix it again as good as before, otherwise thou injurest the teamster to benefit thyself, and he will curse thee every time he passeth. Amen.

AND LASTLY.—Thou shalt be particular not to leave thy bundle at thy boarding-house, without making satisfactory arrangements with the man whose grub thou hast eaten, for it is far better to have a good name than much riches.

APPENDIXVI.

BUSINESSES.

Contained herein are Nelson Point businesses, their earliest date of operation, and if known the location and type of enterprise.

Abbot House (1854) .... Nelson Pt .... gambling
Adams & Co. (see Frank Everts & Co.)
Barnett, R.I. (1854) .... Nelson Pt .... attorney
Bath, Powers, & Co. (1854) .... Nelson Pt .... bowling-alley
Bryant Store (1854-59) .... Independence Bar .... liquor
Captain Cunningham & Co. (1858) .... Nelson Pt .... store-meals
Carr & Co. Store (1862) .... Nelson Pt .... general merchandise
Charlie's Store (Chinese Charlie) (1863)
Cross's Blacksmith Shop (1855) .... Nelson Pt.
Davis & Co. Store (1855) .... Independence Bar
Fegans Saloon (1854) .... Nelson Pt.
Frank Everts & Co. (Adams & Co.) (1854) .... Nelson Pt .... bank
Gilson Boarding House (1853) .... Henpeck Flat
Hambly's Grocery and Liquor Store (1854) .... Nelson Cr.
Hambly's Boarding House (1853) .... Nelson Cr.
Haun & Co. (1855) .... Nelson Cr .... store and cookhouse
Hill, M.D. (1855) .... Nelson Cr .... doctor
Hospital (1854) .... Nelson Cr.
Hundley (1857) .... Nelson Cr .... attorney
Joe Ficklin & Co. (1856) .... Nelson Cr .... stable
John S. Root & Co. (1854) .... Nelson Cr .... merchandise-grocery
Koots & Co. (1854) .... Nelson Cr .... general merchandise
Lewis & Ellis Store (1854) .... Nelson Cr.
Lewis & Roots (1854) .... Independence Bar .... liquor
Main, McDonald, & Co.(1854) .... Independ. B .... merchandise-liquor
Mandonald & Co. Restaurant (1854)(formerly the Gilson House)
McBeth Store (1854) .... Meeker's Flat .... merchandise-liquor
Meyers & Co. (1854) .... Independence Bar .... Liquor
Meyers & Ritchie (1865) .... Nelson Pt .... general merchandise
Meyers & Richards Restraurant (1862) .... Nelson Pt.
Nelson House (see Thompson House)
Nelson Point Store (see Thompson House)
Parker's Store (1854) .... Nelson Pt .... merchandise-liquor
Pauly Hotel (1875) .... Nelson (Lindsey) Pt .... hotel-store
Picketts Store (1854) .... Nelson Cr.

Plumas Saloon (1854)	Nelson Pt.	
Pyles Stable (1862)	Nelson Pt.	
Pyles' Butcher Shop (1862)	Nelson Pt.	
S. Aldridge & Co. (1854)	Nelson Cr	merchandise, dram-shop
Sherwin's Store (1854)	Independence Bar	merchandise-grocery
Snell & Co. (1854)	Nelson Cr	general merchandise
Sylvester Meyers Store (1866).	Nelson Pt	general merchandise
Thompson House (1851)	Nelson Pt	merchandise-tavern-hotel
Thompson Store (1850)	Nelson Pt	general merchandise
Thompson Stable (1851)	Nelson Pt. & Illinois Ranch	
Tilley Trading Post (see Wisconsin & Co. Trading Post)		
Timberman's Store (1854-59)	Nelson Pt	general merchandise
Vaughn, M.D. (1853)	Nelson Cr	doctor
Vegetable Store (1854)	Nelson Pt.	
Wisconson & Co. Trading Post (1850)	Meeker Flat	
Yank's Saloon (1854)	Nelson Cr.	
Zimberman's Store (1854)	Nelson Pt	merchandise-liquor

APPENDIX VII.

NELSONITES.

Contained herein is a list of those that visited, lived or worked at Nelson Creek. Because spellings of the same name varied from one source to another and occasionally the same author spelled a name several different ways, multiple spellings are included. The year cited is that of the earliest documented activity. Whenever possible a brief biographical entry follows for identification purposes.

Abbey—(1850) partner in the Wisconsin Co.

Abbot, ?—(1854) operated the Abbot House

Aldrich, Sylas—(1853) vigilance committee member/operated S. Aldrich & Co. Store in 1854

Alford—(1857)

Allen, Charley—(1854)

Allen, S.—(1864)

Allens, C.—(1855)

Appleby, W.—(1854)

Arnold, Limon—(1853) died accidentally in 1853

Asberry, George W. (Ashbury)—(1854) gambler from Kentucky who destroyed a Nelson Point vegetable store

Ballou, Sam (Baloo)—(1856) surveyor who in 1866 was a partner in Stevens, Ballou & Co./also mined with the "Duke"

Barnett, R.I.—(1854) miner and attorney

Bass, ?—(1853)

Bass, Steve—(1855) juryman in the Haun-Lloyd dispute

Bass, Richard & Family—(1855)

Bass, John—(1855) juryman in the Haun-Lloyd dispute

Bath, A.L.—(1854) partner in Bath, Powers & Co.

Batterton, ? (Batterson)—(1850) co-discoverer of Nelson Creek

Baxter, John—(1853) vigilance committee member/acquitted of the murder of Taffe in 1854

Baxter, ?—(1854)

Beach, ?—(1855) juryman in the Haun-Lloyd dispute

Beach, Mrs.—(1854)

Beaughan, Lindsey—(1905) attended the Nelson Point School

Beck, Christ—(1863)

Benomy, Dick—(1854) died in a mining accident in 1854

Berg, John Gustav—(1884) native of Sweden/mail contractor

Berg, Charles—(1900)

Berk, Christopher—(1865) worked the "old Saw-mill claim"

Berryhill, Dave—(1854)

Bevilacqua, John—(1879) mined the Red Rose drift mine and the Fairbanks Consolidated hydraulic operation

Blackington, Mr.—(1855)

Blair, Samuel Parker—(1863) native of Missouri/partner with Bushman in 1863/registered a miner in 1884 at the age of 69

Blodgett, Miss—(1855) gave a concert at Nelson Point

Blosser, W.H.—(1854)
Bohner, ?—(1868) partners with Lee, Kohr & Francis
Boland, Pat—(1914) mail contractor
Bonner, Rev.—(1855) preached at the Sherwin House
Bonner, ?—(1865) partner in Bonner, Hardy & Co.
Bonner, Squire Thomas D.—(mid-1850's)
Boot, ?—(1855) partner with Freer
Boss, Edwin Harvey—(1884) native of New York
Bowen, Mr.—(1857) debate club member
Bowers, ?—(1854) operated Bowers & Co. hydraulic operation/partner with Smith, Snuffin & Phelps at Independence Bar
Bradley, ?—(1853)
Braun, August—(1852) from Memel/ worked on Rothrock's Nelson Creek claim in 1852
Bray, Lewis H.—(1857)
Bray, Hanibal—(1854) mined with Haun
Bray, Henry—(1857)
Bray, ?—(1855) mentioned in the Haun diary
Bray, H.—(1856)
Brewster, S.T.—(1858) operated a saddle-train
Bridger, Edmund—(1853) died in the Nelson Creek small pox outbreak in 1854
Brinegar, Andrew J—(1851) from near Mineral Point, Wisconsin, who died November 4, 1851, at the age of 27 / buried in the graveyard at Independence Bar
Brooks, ?—(1856)
Brown, H.—(1853) died in the ill-fated winter of '53 escape
Brown, Jesse—(1864) operated Jesse Brown & Co.
Brown, Sam & Son—(1855)
Bryan, Asa—(1855) juryman in the Haun-Lloyd dispute
Bryant, C.—(1854-59) operated a liquor store at Independence Bar
Buckbee, Mr.—(1854)
Bucklin, ?—(1854) gambler who destroyed a Nelson Point vegetable store
Buck—(1854) gambler
Bull, John—(1854) miner hired by Haun
Bummer, Hezekiah—(1868)
Burr, "Chinaman Squires"—(1863)
Bushman, David—(1863) partner with Blair
Bushman, ?—(1864) partner in Johnson, Bushman & Co. at Independence Bar
Butts, ?—(1856)
Byers, J.T.—(1856) butcher
Callich, ?—(1864)
Carpenter, William—(1854) boarded with Hambly
Carr, Dr.—(1863)
Carroll, ?—(1856)
Cate, D.R.—(1855)
Cavanaugh, Edward—(1884) native of Ireland
Chadwick, S.—(1855)
Childress, Henry H.—(1884) native of California/teacher
Chinese Charlie—(1863) operated Charlie's Store
Christendorff, Charley—(1850)
Christendorff, Carl—(1850) illustrated *The California Gold Rush Diary of a German Sailor*
Clark, ?—(1855) Haun's neighbor
Clark, Nat—(1855)
Clay, Cash—(1855)
Cleaveland, Miss—(1856)

Clownie, Mr.(Clowney)—(1853) appointed judge in the Gilson-Wilson trial
Coleman, Willis B.—(1884) native of Pennsylvania/carpenter
Collier, Mrs.—(1855) cooked and washed for hire
Collier, Mr.—(1855) juryman in the Haun-Lloyd dispute
Collier Family—(1855)
Compton, Tim—(1856)
Compton, Lem—(1856)
Conner, R.—(1854)
Cook, J.—(1855) juryman in the Haun-Lloyd dispute
Cook, ?—(1854) hired laborer on Haun's claim
Cooksey, Edward—(1864) infant son of Frances Cooksey presumably born at Nelson Creek in 1864
Cooksey, James—(1854) mined until 1867 when he left for Sierra County
Cooksey, Frances E.—(1862) married James Cooksey in 1862 and lived on Nelson Creek with her husband until 1867
Cooksey, William—(1866) infant son of Frances Cooksey presumably born at Nelson Creek in 1866
Cooksey, George—(1867) infant son of Frances Cooksey born in 1867 presumably at Nelson Creek
Cooleyham, John—(1857) debate club member
Cooper, S.J.—(1854)
Cooper, ?—(1856)
Copeland, ?—(1865) operated Copeland & Co.
Cornelison, ?—(1854)
Corson, R.—(1855)
Cotter, Robert—(1924)
Cotton, Aylett R.—(1850) led the Cotton prospecting party/wrote *Across the Plains to California in 1849 And After: An Autobiography*
Crampton, William Alexander—(1892) native of Alabama
Crook, David—(1864)
Cross, Horatio—(1854)
Cross, Lewis—(1854)
Cross, L.—(1853) vigilance committee member
Crumpton, J.P.—(1924) miner whose car burned in the fire of '24
Cunningham, Capt. Noble C.—(1858) restraunter captain of the Plumas Rangers/returned to Missouri in 1868
Curley, Pat (Culey)—(1854) Independence Bar
Cutler, Elihu—(1900)
Davenport, ?—(1853) mined with Edwards
Davis, "Big John"—(1855)
Davis, Jason I.—(1857)
Davis, Joe—(1855) ran a pack-train
Davis, ?—(1855) operated Davis & Co. Store
Davison, A.—(1854)
Dean, Mrs.—(1856)
Dean, W.S.—(1855) operated a saddle-train
Deavenport, William—(1857) one of Haun's mining partners
Delahunty Brothers—(1902) mail contractor
Delano, Alonzo—(1850) wrote *Life on the Plains and Among the Diggings* in 1854 which was later reprinted as *Across the Plains and Among the Diggings* in 1936
Delerro, Gus—(1863) partner with Gildersleeve
Delevan, Mrs.—(1855)
Delevan, Joe—(1857)

Dever, Mrs.—(1854) washed for hire
Dickson, W.R.(Dixon?)—(1854) he is likely the Dixon for which Dixon Creek was named
Dilks, William L.—(1866) native of Pennsylvania
Dobb, ?—(1856)
Dobson, I.—(1853)
Dobson, ?—(1854) one of Haun's partners
Dobson, Pat—(1854)
Doosley, ?—(1854)
Douglap, ?—(1854)
Douglas, Charles—(1854) died in a mining accident in 1854
Doven, J.K.—(1857)
Dover, ?—(1858)
Dow, John J.—(sometime after 1857) mined at Nelson Point
Dow, John L.—(1900)
Dow, Miss Belle—(?) died of the "Great Destroyer" in 1904 at the age of 23. Daughter of Mr. & Mrs. John Dow
Dow, Mary—(1905) presumably the daughter of John L. Dow/attended the Nelson Point School
Dow, Mildred—(1905) presumably the daughter of John L. Dow/attended the Nelson Point School
Dow, Ward—(1905) presumably the son of John L. Dow/attended the Nelson Point School
Downie, Patrick—(1857)
Downie—(1854) held a concert at Nelson Point
Drake, ?—(1854)
Drumgold, James Henry—(1892) native of New York
Duesler, Mrs. (Dousler)—(1855) worked at the Roots & Lewis Store
Duesler, Mr.—(1855) friend of Haun
Duke—(1866) mined with Sam Ballou
Durgin, W.—(1855)
Durrill, Orrin—(1884) native of Maine
Dutch Charlie—(1856) mined on Nelson Creek at Rocky Bar
Eaves, Tom (Eves)—(1855) mined with Haun
Eddlbudel, John C. (Eddelbuttel, Johann Conrad)—(1884) native of Hanover, Germany
Edwards, Ebenezer—(1857)
Edwards, Mr. & Mrs.—(1855)
Edwards, Jane—(1855) Edwards's wife
Edwards, Earl—(1857) Alford's uncle
Edwards, James E.—(1855) native of England/mined Nelson Creek and later operated the Plumas House
Edwards, William A.Z.—(1853) wrote the Edwards Diaries
Egbert, Harvey W.,Sr.—(?) attended the Nelson Point School at the turn of the century and bought the Willow Ranch in 1928
Eggbert, Harvey William (Egbert)—(1902) worked the Nelson Mine
Egbert, Henry William—(1884) native of Ohio
Ellerman, ?—(1866) mined with Gohda (Gohile?)
Ellerman, Charles A.—(1875) native of Germany/registered as a miner in 1875 at the age of 28
Emmons, ?—(1863)
Ettlebottle, Louise—(1905) attended the Nelson Point School
Everts, Frank—(1850-1) banker and express agent
Everts, Henry—(1850-1) Frank's brother/expressman
Fadder, Herman—(1863)
Farleigh, Mr.—(1854)
Farley, M.H.—(1853) vigilance committee member
Fegan, J.J. (Fagan)—(1853) operated Fegan's saloon

Fell, Edward S.—(1857)
Fell, G.W.—(1855)
Felt, G.—(1855)
Ferguson, Charles D.—(1850) wrote *The Experiences of a 49er During a Third of a Century in the Gold Fields* in 1888
Ficklin, Joe—(1856) operated a stable
Fields, Edwin Augustus—(1884) native of Massachusetts
Fitzgerald, D.—(1856)
Flourney, Robert Simpson—(1852) 49er who mined at Nelson Point and Rich Bar for many years
Folger, Miss—(1855)
Foots, ?—(1856) partner with Walker in a saddle-train
Foster, "Spanish Bill"—(1862)
Fowler, Simeon (Simon)—(1856) found a dead chinaman on the Washington Hill trail
Fox, Mrs.—(1855) mother of Frank S. Fox
France, William—(1875) native of New York
Francis—(1868) partners with Lee, Kohr & Bohner
Freer, ?—(1854) partner with Boot in 1855
Frink, Billy—(1862) mined with Doc Vaughn
Frisk, Peter—(1892) native of Sweden
Funk, B.—(1857)
Gallagher, Patrick—(1855)
Ganeshki, Jerry—(1919)
Garbani, Gaspero—(1884) native of Switzerland
Garbani, Guiseppi—(1884) native of Switzerland
Gentry, J.P.—(1854) minister
Gildersleeve, ?—(1854)
Giles, ?—(1865) partner in Mitchell, Giles & Co.
Gilleland, John—(1875) native of Pennsylvania
Gilson, Mr. & Mrs.—(1853) operated a boarding house at Henpeck Flat
Givens, ?—(1854)
Glass, ?—(1854)
Goble, Frank—(1856)
Gohda, ?—(1866) mined with Ellerman
Gohile, ?—(1866) (Gohile and Gohda are likely the same person) partner with Ellerman
Goodenow, Miss(Goodnow)—(1854) gave a concert at Nelson Point
Goodshall, ?—(1854) "an old dutchman"
Goodspeed, Walter—(1853) died in the ill-fated winter of '53 escape
Gorson, ?—(1855) mined with Larison
Gould, Arthur T.—(1918) mail contractor
Grace, Jim—(1856) operated the Gibsonville-Nelson Point express
Grant, Mr.—(1855)
Graves, Misses—(1855) gave a concert at Nelson Point
Gray, Antonio—(1854) boarded Edwards
Greenwood, Noah—(1854)
Greeny, Joe—(1855)
Griffith, George—(1850)
Gung, A.—(1863) Chinese
Gunther, Fritz—(1852) Rothrock's brother-in-law/worked on Rothrock's Nelson Creek claim in 1852
Haffenger, Henry Bennett—(1884) native of New York
Haislip, ?—(1850)
Hambly, William—(1852) David Hambly's brother
Hambly, David—(1852)
Hambly, Ophelia Anne Dee—(1854) daughter of Hambly
Hambly, Mrs.—(1853) David's wife
Hammerman, George H.G.—(1884) native of New York

Hampton, ?—(1853)
Hannon, ?—(1856) expressman
Hardin, ?—(1854) operated Hardin & Co.
Hardy, ?—(1865) partner in Bonner, Hardy & Co.
Hardy, Captain—(1853) vigilance committee member
Hargrave, Abe—(1853) vigilance committee member
Hargrave, Joseph P.—(1853) appointed Nelson Creek recorder March 19, 1853
Harkins, Patrick—(1884) native of Ireland
Harmon, ?—(1864) partner in Harmon, Oliver & Co. at Henpeck City (Flat)
Harms, John—(mid-1860's)
Harriett, Miss—(1855)
Harris, Orin C.—(1884) native of New York
Hatch, S.B.—(1850) partner in the Wisconsin Co.
Haun, Pauline—(1855)
Haun, Martha—(1855) Haun's wife
Haun, James Humphrey—(1853) wrote the Haun Diary
Haun, Lizzy—(1855) Haun's niece
Hawkins (Hawkings)—(1854) one of Haun's partners
Heath, Judge—(1855) mined with Haun
Henry, John—(1856)
Henry, Mrs. John—(1856)
Hermann—(1863)
Hill, Doc—(1853)
Hill, ?—(1865) operated Hill & Co. at the lower end of Columbia Flat on the Middle Fork
Hireman, ?—(1856)
Hopkins, ?—(1854) cousin of Hawkins
Hough, Bill—(1850)
Houk, David—(1900)
Howard, James—(1863) partner with Reede
Howe, Sam—(1863)
Howland, Joseph—(1857) debate club member
Hubbell, Eura Sherman—(1884) native of Canada/saloonkeeper
Hughs, ?—(1924) miner whose car burned in the fire of '24
Humphreys, John—(1854)
Hundley, ?—(1857) attorney
Hunley, ? (Hundley)—(1854) attorney
Huntington, ?—(1857)
Ivey—(1850) partner in the Wisconsin Co.
Jackson, ?—(1866) mined with Jolly
Jackson, Andrew—(1875) native of Ohio
James, John—(1855)
James, William J.—(1884) native of Pennsylvania
James—(1924) miner whose car burned in the fire of '24
Jarrett Brothers—(1857)
Jennings, G.—(1855)
Jennings, Isaac—(1855)
Jewett, Mr.—(1850) from Massachusetts
Johnson, ?—(1864) partner in Johnson, Bushman & Co. at Independence Bar
Johnson, ?—(1863) operated Johnson & Co.
Johnson, A.—(1854)
Johnson, ?—(1856) mined at Independence Bar
Johnson, F.—(1884) native of Massachusetts
Johnson, Frank Even—(1900) worked the Nelson Mine in 1902
Jolly, Thomas—(1866) native of Illinois/mined with Jackson in 1866/registered as a miner in 1884 at the age of 42
Jones, A.—(1857)
Jones, Austin "C"—(1900)
Joulan—(1864)
Joy, Charles Edward—(1900)
Judkins, ?—(1853)

Kamke, Charles—(1852) from Dantzig/ worked on Rothrock's Nelson Creek claim in 1852.
Kane, ?—(1866) mined with Reed
Kain, Frank—(1875) native of Ireland
Kaneckdate, Jerry—(early 1900's) arrived at Nelson Point when he was seventeen
Keene, Lem—(1856)
Keene, Edward—(1854)
Kellogg, ?—(1855)
Kelly, Andrew—(1850) partner in the Wisconsin Co.
Kelly, Ripley C.—(1850) partner in the Wisconsin Co.
Kepner, John Kline—(1852)
Kimball, ?—(1855)
Kinman, ?—(1855) juryman in Haun-Lloyd dispute
Kitto, Thomas—(about 1892) mined for 15 years in and around Nelson Point
Kline, George—(1854)
Knock, F.—(1867)
Knowles, William H.—(1868) native of New York/snow-shoe expressman
Koher, ? (Kohr)—(1866) partner in Koher, Lee & Co. in 1866/partner with Lee, Francis & Bohner in 1868
Koots, ?—(1854) operated Koots & Co.
Kyler, ?—(1856) mined with Haun
Labadie, Henry Warren—(1900)
Lacy, Mrs.—(1856)
Lane, ?—(1854) suspected of killing a chinaman
Lanno, August—(1892) native of France
Larison, James H. (Larrison)—(1853)
Lasswele, ?—(1863) mined with Mitchele
Laufman, Cyrus—(1850) native of Pennsylvania/he arrived at Nelson Creek in April as one of the Gold Lake searchers/mined Nelson Creek and the South Fork of the Feather until the winter of 1851
Lawrance (Lawrence)—(1854) mined with Haun in 1855
Lawrence, James—(1850) miner from Vermont who was murdered with a hatchet
Lawson, Peter—(1884) native of Sweden/sawyer
Lecouvreur, Frank—(1852) wrote *From East Prussia to the Golden Gate*/worked on Rothrock's Nelson Creek claim in 1852
Lee, ?—(1866) partner in Koher, Lee & Co./partner with Kohr, Francis & Bohner in 1868
Lemieux, Joseph—(1906) mail contractor
Levi, ?—(1854) mined with Haun
Lewis, James C.—(1854) lost to John Harbison the Plumas Co. Clerk election in the fall of 1854
Lewis, David—(1854) mined Wiley's Flat
Limberman, William (Limmerman)—(1854) merchant who operated Oldrige & Limmerman
Linch, ?—(1854) 49er from Kentucky
Lindley, Mrs.—(1856)
Lindsey, Edwin Galloway—(1892) native of Wisconsin/worked the Nelson Mine in 1902
Lindsey, Edward G.—(late 1800's)
Lindsey, Mrs. Edward G.—(late 1800's) native of Nelson Point/daughter of J.H. Schubert/resident of Nelson Point as late as 1904
Lindsey, George H.—(1905) presumably the son of Mr. and Mrs. Edward G. Lindsey/attended the Nelson Point School
Linscott, J.M.—(1854) mined at Wiley's Flat
Lloyd, John—(1854) one of Haun's partner
Long, A.—(1862) chinaman who lived at Columbia Flat

Long, William—(1884) native of Tennessee
Lovejoy, Sherwin—(1855)
Lowery, D.B.—(1890) mail contractor
Luther, Wesley—(1850) miner from Vermont who was murdered with a hatchet
Lyons, W.—(1863) partner with Tack
Madden, Osbourne (Osborn) & Family—(1854)
Madden, Jason—(1854) partner with Osbourne in a bar at Nelson Point
Maguire, ?—(1865) operated Maguire & Co.
Maguire, Patrick—(?) mined the East Branch of Nelson Creek/died in the early 1890's in his seventies/very likely the same man who operated Maguire & Co.
Mandeville, Frank Charles—(1900) worked the Nelson Mine in 1902/by 1912 he had commenced farming in Quincy and had married Annie Sidona Mandeville
Mandeville, Bessie—(1905) daughter of Frank and Annie/attended the Nelson Point School
Mandeville, Ida—(1905) child of Frank and Annie/attended the Nelson Point School
Mandeville, Margaret—(1905) daughter of Frank and Annie/attended the Nelson Point School
Main, ?—(1854) partner in Main, McDonald & Co. Store
Mandonald (Maindonald), Fred—(1853) bought Gilson's Boarding House/operated general merchandise stores at Nelsonville and Independence Bar
Marken, ?—(1865) operated Marken & Co. at Fiddler's Flat
Marks, Mr.—(1850)
Marshall, D.—(1855)
Martin the Hungarian—(1857) old resident of Nelson Creek
McBeth, John—(1854) operated a store at Meeker's Flat
McCartney, Bill—(1856) operated the Gibsonville-Nelson Point express
McClure, ?—operated a saddle train
McCummings, ?—(1857)
McDonald, Fred (Maindonald or Mandonald)—(1853) vigilance committee member
McEwer, ?—(1864) operated McEwer & Co.
McFall, ?—(1854) "old man"
McGuffin, ?—(1855)
McGuire, Redmond—(1875) native of Ireland
McGuire, Ed(McGuyre)—(1853) operated McGuire & Co./killed in 1856 when a bank caved in
McHatton, Green—(1854) 49er from Illinois
McIntosh, Alex—(1855)
McKenzie, ?—(1855)
McKitrick, ?—(1855)
McNabb, ?—(1856)
McNabb, James H.(McNable)—(1853) referred to as Squire McNabb in 1854/was referred to as a minister in 1857
Mechling, Jacob J.—(1884) native of Pennsylvania/tinsmith
Mechling, Joseph J.—(prior to 1890) lived in the vicinity of Nelson Point
Meeker, Warner (Warren)—(1850) Meeker Flat named for he and his family/partner in the Wisconsin Co.
Meeker, Jonathan—(1853) vigilance committee member
Merrill, Mr.—(1855)
Metcalf, William, Sr.—(1865) native of Ireland/supervisor of the Blue Gravel Mine

Metcalf, Elizabeth Kelly—(1872) William Metcalf, Sr.'s wife
Metcalf, William H.—(1875) son of William and Elizabeth/miner and sawmill operator at Nelson Point/ married Annie Elizabeth Otis
Metcalf, Annie Elizabeth—(about 1900) native of Ireland/married William H. Metcalf at the age of 22
Metcalf, Edward—(1871) native of California/registered as a miner in 1892 at the age of 21/worked the Nelson Mine in 1902
Metcalf, Samuel—(1884) native of Ireland/presumably William's brother
Metcalf, Mrs. Samuel—(late 1800's) native of Ireland
Metcalf, Peter—(1884) native of Wisconsin
Meyers, ?—(1850) partner in the Wisconsin Co.
Milligan, ?—(1854) stabbed Davis
Milling, S.J.—(1854)
Millington, ?—(1854) operated a dancing school
Mitchele, ?—(1863) mined with Lasswele
Mitchell, ?—(1865) partner in Mitchell, Giles & Co.
Moller, Peter Christian—(1884) native of Prussia
Moncur, O.—(1924) last tenant of the Pauly Hotel
Moo, A.—(1856) storeowner
Moore, ?—(1857)
Morricey, ?—(1855)
Morris, Charles—(1855) operated Charles Morris & Co.
Morton, Mr.—(1854) held a concert at Nelson Point
Muicken, Diederrich (Murcken or Murken)—(1863) partner with Reinke in 1865/mined with Schubert in 1866
Muma, ?—(1856) expressman
Murray, Mr.—(1850)
Myers, Sylvester—(1855) native of New York/merchant
Myrtle, Lois—(1890's) school teacher
Mysenheimer, ?—(1857)
Nabb, J.W.H.—(1857) debate club member
Naylon, W.F.S.—(1855)
Neal, Tom—(1910) mail contractor
Neal, Dept. Sheriff—(1854)
Nelson, ?—(1850) co-discoverer of Nelson Creek
Nescio, Nomen—(1860's) was published regularly in the Union as the "Letter From Nelson Point."
Nutting, Alden J.—(1850) from Westford, Mass./came with Webster from Boston and established a claim on the Middle Fork approximately 30 miles above Bidwell's Bar prior to he and Webster moving up to Nelson Creek
O'Grady, ?—(1856)
Oikle, Tessie—(1920's) lived alone in a cabin on Nelson Creek during the 1920's and 1930's/was in her 60's in 1973
O'Brien, ?—(1857)
O'Bryan, Edward—(1884) native of Ireland
O'Rourke, L.B.—(1922) mail contractor
Old Frank—(1856)
Olson, Magnus—(1900)
Olias, Mr.—(1852) mined on the North Fork at the confluence of Nelson Creek
Oliver, Richard—(1863) native of Canada/operated Richard Oliver & Co./partner in Harmon, Oliver & Co.and in Giles, Oliver & Co. at French Hill in 1864
Oliver, George—(1875) native of Michigan

Ossdyke, S.H.—(1856)
Page, A.L.(Paige)—(1853) vigilance committee member/suspected of killing a chinaman in 1854
Parker, W.W.—(1854) operated a store at Nelson Point
Parker M.—(1853) vigilance committee member
Parmeter, J.—(1855)
Parsons, Mr.—(1854) succeeded Hambly as hospital operator
Pauly, Ben—(late 1800's) had the mail contract in 1916-18 on the Quincy-La Porte stage
Pauly, Henry Robert—(late 1800's) threw the male teacher out the door when he was 15 at the Nelson Point school
Pauly, Erle—(1909) grandson of Nicholas
Pauly, Catherine (Katherine)—(about 1875) wife of Nicholas Pauly
Pauly, Nicholas Oscar—(about 1875)
Pauly, Julius E.—(1875) (son of Nicholas Pauly) worked in the family store and freight business
Pauly, Bertha—(late 1800's) married Judge Moncur about 1915
Pauly, Oscar Nicholas—(1877) native of Nelson Point/son of Catherine/stage driver/later operated a saloon in Oroville
Payne, Col.—(1854)
Pearce, Mrs.—(mid-1850's)
Peck Family—(1854) Mr. Peck died in 1854
Peel, John J.L.—(1851) from Tennessee/mined near Nelson Point until 1856/he later settled in Truckee
Perkins, ?—(1857)
Phelps, ?—(1854) partner with Bowers, Smith & Snuffin
Phillips, William—(1853) died in the ill-fated winter of '53 escape
Pickett, ?—(1854) operated Pickett's Store
Pike, ?—(1855)
Pike, Jim—(1854)
Pinkham, Prof.—(1854)
Poor, George—(1855) juryman in the Haun-Lloyd dispute
Post, Leroy J., Sr.—(1916) delivered mail to the Pauly Hotel
Potts, C.—(1855)
Powers, ?—(1854) partner in Bath, Powers & Co.
Preston—(1854)
Prideaux, William—(1855)
Prideaux, Russell—(1855)
Prisbie, Mr. and Mrs.—(1854) gave a concert at Nelson Point
Prisk, John P.—(1855)
Pritchard, Jason—(1857)
Pyles, ?—(1862) operated a butcher shop and stable
Quigley, Andrew Jackson—(late 1880's) bridge builder
Rains, Bill—(1855) "Rains" in the Haun Diary
Ransmire, Mr.—(early 1850's) filed a lawsuit over a claim that was heard by Squire Bonner
Ransom, L.W.—(1850) reported the hatchet murders to the *Maryville Herald*
Reed, ?—(1866) mined with Kane
Reede, ?—(1863) partner with Howard
Reinhard, Mr.—(1852) worked on Rothrock's Nelson Creek mine in 1852
Reinke, William F.—(1864) partner with Murcken in 1865
Remington, Edward Willis (Remmington)—(1892) native of California
Reynolds, G.—(1856)
Rhinsmeyer, Adolph—(1859) German

miner who was granted citizenship in 1859
Richards, Job—(1855)
Richards, Jacob—(1854)
Richards, J.—(1862) partners with Meyers in a Nelson Point restaurant
Riddle, Tom—(1855) mined the Scotch Co.'s Creek Claim
Riehl, Mr.—(1856)
Riehl, Miss—(1856)
Riehl, Mrs.—(1856)
Riffal, A.—(1864)
Ritchie, John A.—(1857)
Ritterbush Brothers—(1855)
Robinson Family—(1857)
Robinson, Susan—(1857) unmarried member of the Robinson family
Roister, ? (Royster)—(1855) mined with Haun
Root, John S.—(1853) operated John S. Root Store/vigilance committee member
Root, C.M.—(?)
Roots, William—(1857)
Rose, John—(1850) member of the Marks prospecting party
Rosite, ?—(1855)
Rothrock, Mr.—(1852) teamster and miner from Nelson Creek/drove a team of 8 oxen to the Wisconsin House
Rubleneck, Francis—(1875) native of Austria/hotelkeeper at the head of Nelson Creek
Runnells, George—(1855) juryman in the Haun-Lloyd dispute
Rush, Samual—(1856) Independence Bar
Russell, Gen.—(1853)
Russell, Lawson Myrick—(1900)
Russle, Miss—(1856)
Sager Family—(1857)
Schmetzer, Fritz—(1852) worked on Rothrock's Nelson Creek mine in 1852
Schmidt, John—(1863)
Schubert, John Henry (Shubert)—(early 1860's) native of Prussia/mined with Murken in 1866/bought the Willow Ranch in 1874/lumberman in 1875/ died in 1904 at the age of 68
Schubert, G. Henry—(late 1800's) native of Nelson Point who as an adult operated a studio in La Porte and photographed the region/son of John Henry Schubert
Schubert, Annie Sidona—(late 1800's) native of Nelson Point/daughter of John Henry Schubert/married Frank Charles Mandeville
Scott, I.—(1855)
Seatherman, ?—(1855) fought Parmeter at Independence Bar
Seeliger, ?—(1857)
Shaffer, G.W.—(1857)
Shann, Mr.—(1855)
Shannon, R.—(1854) mined the India Rubber Claim
Shargur, H.L.—(1856)
Sharp, Sheriff—(1854)
Shaw, ?—(1854) one of Haun's partners
Sherman, ?—(1865) partner with Winders
Sherwin, J.L.C.—(1850)
Sherwin, P.M.—(1856) mined and sold water to the Rocky Bar mines
Shores, Miss—(1856)
Shults, James & Family (Shultz)—(1855) miner hired by Haun
Sidoua, Dyer—(1905) attended the Nelson Point School
Shultz, Dave—(1856) "Shultz" in the Haun Diary
Silleger, ?—(1857)
Skinner, E.S.—(1857)
Smale, John—(1854)
Small, Mr.—(1854)

Smith, Jas—(1857)
Smith, Harry—(1855)
Smith, Allen—(1864)
Smith, ?—(1854) partners with Bowers, Snuffin & Phelps at Independence Bar
Smith, J.H.—(1855) juryman in Haun-Lloyd dispute
Smith, John—(1863)
Smith, G.E.—(1862) operated a pack-train from Virginia City
Smith, H.E.—(1853) mined with Edwards
Snell, ?—(1853) expressman/operated the Snell & Co. Store in 1854
Sniscott, D.—(1853) mined the Island Bar Claim
Snuffin, ?—(1854) partner with Bowers, Smith & Phelps
Snyder, ?—(1857)
Sockum, ?—(1856)
Spencer, O.H.—(1853) poll clerk in first county election
Squires, Eliakim—(1893) customer of the Pauly Hotel
St. Clair, Miss—(1855) traveling magician
Stanhouse Brothers (Rob, Alex & Richard)—(1855)
Steinhouser, Richard—(1857)
Steinhouser, Robert—(1857)
Sterling ("the Sterlings")—(1856) mined at Rocky Bar
Sterling, Edward H.(W.)—(1853) vigilance committee member/debate club member in 1857
Stevens, B.J.—(1855) attempted suicide
Stevens, B.B.—(1856) debate club member in 1857
Stevens, ?—(1866) partner in Stevens, Ballou & Co.
Strang, M.—(1853)
Streeter, M.R.—(1855) poll tax collector
Swain Family—(1855)
Swain, Mary—(1855) member of the Swain family
Sweeten, ?—(1857)
Tack, ?—(1863) partner with Lyons
Taffe, Peter—(1853) vigilance committee member/stabbed to death allegedly by Baxter in 1854
Taft, D.R.—(1853) mined with Edwards
Taylor, T.—(1854)
Taylor, Jobe T.—(1852)
Taylor, George—(1856)
Teft, ?—(1855)
Tell, ?—(1856)
Terwilleger, ?—(1856) John Terwilleger's son
Terwilleger, John (Terwilliger)—(1851) partner in the Thompson House
Terry, Francis Collins—(1900)
Thomas, David Philip—(1900)
Thomas, N.E.—(1905) school teacher
Thomas, J.M. (alias Pike County Jim)—(1856)
Thomason & Brother—(1854)
Thompson, James M.—(?) made a pencil sketch of Nelson Point while visiting his cousin, John Thompson, which was later used by his sister to create the painting now on display at the Plumas County Museum
Thompson, John Wesley—(1850)
Thorpe, ?—(1864) operated Thorpe & Co.
Tilly, George—(1850) partner in the Wisconsin Co.
Tilly, W.J.—(1850) operated the Wiconsin Co.'s trading post/by the 1880's he relocated to San Francisco
Timberman, William—(1857)
Titus, ?—(1855) juryman in the Haun-Lloyd dispute
Toppafs, James—(1854) shot by Glass in 1854

Toy—(1850) partner in the Wisconsin Co.
Trembath, Thomas—(1884) native of England
Tribe, ?—(1854) suspected of killing a chinaman
Trotter, Charlie—(1862)
Trumble, James—(1854) drowned in 1854
Tucker, William Morris—(1900)
Turner, Alex—(1850) partner in the Wisconsin Co.
Turner, Ed. J.—(1924)
Turvedrop, Mr.—(1862)
Tyler, G.—(1902) worked the Nelson Mine
Vaughn, Doc—(1850)
Vaughn, Mrs.—(1855) Doc Vaughn's wife
Voice, Louis—(1855)
Wade, Adam—(1864)
Walker, Frank—(1855) died accidentally in 1855
Walker, ?—(1856) partner with Foots in a saddle-train in 1856
Wan, Quong—(1857)
Ward, ?—(1850) miner from Vermont who survived a hatchett attack
Warley, James—(1850) partner in the Wisconsin Co.
Warley, John—(1850) partner in the Wisconsin Co.
Waterworth, Jason (James) William—(1884) native of Iowa
Webster, Kimball—(1850) wrote *The Goldseekers of '49*
Weiss, J.—(1855)
Westonhaver, ?—(1855)
Wheat, J.—(1855)
Wheeler, Mat—(1856)
Whickizer, J.(Wickiser)—(1855) mined at Independence Bar
White, Charles & Brother—(1855)
White, W.W.—(1854)
White, Henry—(1855) resident of Nelson Point in 1907 when he died at the age of 71
White, Henry—(1875) native of England/registered as a miner in 1875 at the age of 30
White, W.B. & Family—(1857)
White, Elizabeth—(1857) wife of W.B. White
White, Annie E.—(1859) infant daughter of Elizabeth who died in 1860 at the age of 6 months.
White, Edward A.—(1857) Elizabeth's infant son who died in 1857
White, William B.—(1853) prosecutor in the Gilson-Wilson trial
White, ?—(1854) suspected of killing a chinaman
Whitlock, Charles—(1853) appointed clerk in the Gilson- Wilson trial
Whitlock, Charley—(1853)
Whitlock, Maj. James H.—(1853) defended Gilson in the Gilson-Wilson case
Winter, Gough (Guuke) Rudolph—(1875) native of Holland
Whitney, Messrs—(1924) miner whose car burned in the fire of '24
Whitney, J.D.—(1855) geologist
Wiess, Jacob—(1854)
William, France—(1875) native of New York
Williams, ?—(1855)
Williams, Tom—(1853) one of Haun's original partners
Williams, ?—1868 operated the mail contract with Brewster
Wilson, Mr.—(1853) shot by Gilson
Wilson, ?—(1856) expressman
Wilson, R.R.—(1853)

Wilson, C.—(1902) worked the Nelson Mine

Windeler, Adolphus—(1850) wrote *The California Gold Rush Diary of* a *German Sailor*

Winders, Hile—(1865) partner with Sherman

Winters, Jason—(1857)

Winters, Capt. George R.—(1866)

Winters, Gauke (Gough) Rudolph—(1875) native of Holland

Woden, James—(1853) vigilance committee member

Woo, Mi—(1883)

Wooden, James—(1855) vigilance committee member

Yankee Jack—(1855) an old Frenchman

Young, John—(1855) presumably the same John Young appearing in the Nelson Point Lumber Company receipt, 1901 (*see* Chapter XX, Lumber)

Young, I.M.—(1855)

Young, W.—(1854)

Zimberman, ?—(1854) operated Zimberman's Store

APPENDIX VIII.

CHRONOLOGY OF NEWSPAPER PUBLICATION.

Contained herein is a chronology of 19th century newspaper publication in the region surrounding Nelson Point.

Nov. 19, 1853	*The Trumpet*—published semi-weekly at Gibsonville; renamed the *Gibsonville Trumpet & St. Louis News*; later renamed the *Mountain Messenger* and changed to a weekly publication in May of 1854; moved to Rabbit Creek in March of 1856; moved to Downieville, Sierra County, in March of 1864 where it is currently published.
August, 1855	*The Old Mountaineer*—The *Argus, Plumas Democrat,* and the *Fillmore Banner* were all published from the *Old Mountaineer* print shop in Quincy during the election of 1856.
1857-60	*The Plumas Argus* (I)
Nov. 12, 1859	*The Plumas Standard*—published in Quincy; moved to Downieville in 1864.
1860	*The Plumas Argus* (II)—reborn for three months.
Nov. 25, 1861	*The Quincy Union*—moved to La Porte and renamed the *La Porte Union* on Dec. 5, 1868; destroyed by fire in July of 1869.
Aug. 11, 1866	*The Plumas National*—published in Quincy by the Union Party of Plumas County.
May 16, 1892	*The Plumas National/Bulletin*—two papers merged by the Plumas County Publishing Co.

APPENDIX IX.

MINING COMPANIES.

Contained herein is a supplementary list of mining companies that operated on Nelson Creek.

Fell & Co.
Grand Bamboo Co.
Honeycut & Co.
I. Fee & Co.
India Rubber Co.
Irish Co.
Joint, Stock, Co.
Last Hope Co.
Maden & Co.
Overtone & Co.
Poor & Co.
Portugeese Co.
Sardine Co.
Scotch Co.
Stanhouse & Co.
Tweedle Dee Co.
Zark Co.

APPENDIX X.

MINING CLAIMS.

Contained herein is a supplementary list of mining claims that operated on Nelson Creek.

- Deadwood Claim
- Deadwood No. 2 Claim
- Free and Easy Claim
- Grand Bamboo Co. Claim
- Island Bar Claim
- Last Hope
- Michigan Diggings
- Scotch Co.'s Creek Claim

APPENDIX XI.

VOTER REGISTRATION.

Contained herein is a sampling of the registered voters from Nelson Point. Entries from the 1875, 1884 and 1892 Plumas County registers were extracted as they appeared in the original documents. By 1900 voters were categorized under the Nelson Point Precinct and those are presented here in their original form.

GREAT REGISTER OF PLUMAS COUNTY, CAL.
AUGUST, 1875

	Age	Nativity	Occupation	Local Residence
France, William	31	N.Y.	Miner	Nelson Point
Gilleland, John	55	Pennsylvania	Miner	Nelson Point
Jackson, Andrew	32	Ohio	Miner	Nelson Point
Kain, Frank	42	Ireland	Miner	Nelson Point
Knowles, Wm. H.	46	N.Y.	Miner	Nelson Point
McGuire, Redmond	32	Ireland	Miner	Nelson Point
Myers, Sylvester	36	N.Y.	Merchant	Nelson Point
Oliver, Richard	33	Canada	Miner	Nelson Point
Oliver, George	23	Michigan	Miner	Nelson Point
Rubleneck, Francis	42	Austria	Hotel Keeper	Head of Nelson
Schubert, Henry	32	Prussia	Lumberman	Nelson Point
White, Henry	30	England	Miner	Nelson Point
Winter, Gough Rudolph	41	Holland	Miner	Nelson Point
Ellerman, Charles A.	28	Germany	Miner	Nelson Point

GREAT REGISTER OF THE COUNTY OF PLUMAS,
STATE OF CALIFORNIA
OCTOBER 1884.

	Age	Nativity	Occupation	Local Residence
Blair, Samuel Parker	69	Missouri	Miner	Nelson Point
Berg, John Gustav	38	Sweden	Mail Carrier	Nelson Point
Boss, Edwin Harvey	31	N.Y.	Miner	Nelson Point
Cavanaugh, Edward	32	Ireland	Miner	Nelson Point
Coleman, Willis	37	Pennsylvania	Carpenter	Nelson Point
Childress, Henry H.	26	California	Teacher	Nelson Point
Durrill, Orrin	42	Maine	Miner	Nelson Point
Dilks, Wm. L.	49	Pennsylvania	Miner	Nelson Point
Eddlebudel, John C.	48	Hanover	Miner	Nelson Point
Ellerman, A. Charles	32	Germany	Miner	Nelson Point

Egbert, Henry	47	Ohio	Miner	Nelson Point
Fields, Edwin Augustus	45	Massachusetts	Miner	Nelson Point
Garbani, Guiseppi	52	Switzerland	Miner	Nelson Point
Garbani, Gaspero	31	Switzerland	Miner	Nelson Point
Hubbell, Eura Sherman	55	Canada	Saloon Keeper	Nelson Point
Harris, Orin C.	53	N.Y.	Miner	Nelson Point
Harkins, Patrick	48	Ireland	Miner	Nelson Point
Haffenger, Henry	53	N.Y.	Miner	Nelson Point
Hammerman, Geo. H.G.	40	Germany	Miner	Nelson Point
Jolly, Thomas	42	Illinois	Miner	Nelson Point
Johnson, F.	41	Massachusetts	Miner	Nelson Point
James, William J.	42	Pennsylvania	Miner	Nelson Point
Kain, Frank	54	Ireland	Miner	Nelson Point
Lawson, Peter	46	Sweden	Miner	Nelson Point
Long, William	50	Tennessee	Miner	Nelson Point
Metcalf, William	41	Ireland	Miner	Nelson Point
Metcalf, Samuel	47	Ireland	Miner	Nelson Point
Metcalf, Peter	24	Wisconsin	Miner	Nelson point
Moller, Peter Christian	40	Prussia	Miner	Nelson Point
Mechling, Jacob J.	50	Pennsylvania	Tin Smith	Nelson Point
Oliver, Richard	45	Canada	Miner	Nelson Point
Oliver, George	29	Michigan	Miner	Nelson Point
O'Bryan, Edward	30	Ireland	Miner	Nelson Point
Pauly, Nicholas O.	49	Germany	Hotel Keeper	Nelson Point
Schubert, Henry	44	Prussia	Farmer	Nelson Point
Trembath, Thomas	29	England	Miner	Nelson Point
Waterworth, Jas. Wm.	21	Iowa	Miner	Nelson Point
Winters, Guuke Rudolf	51	Holland	Miner	Nelson Point

GREAT REGISTER OF PLUMAS COUNTY
STATE OF CALIFORNIA
1892

	Age	Nativity	Occupation	Local Residence
Boss, Edwin Harvey	38	NY	Miner	Nelson Point
Crampton, William Alexander	70	Alabama	Miner	Nelson Point
Durrell, Orrin	58	Maine	Miner	Nelson Point
Drumgold, James Henry	31	NY	Miner	Nelson Point
Eddelbuttel, Johann Conrad	59	Germany	Miner	Nelson Point
Frisk, Peter	36	Sweden	Miner	Nelson Point
Lindsey, Edward	24	Wisconsin	Miner	Nelson Point
Lawson, Peter	59	Sweden	Sawyer	Nelson Point

Lanno, August	72	France	Miner	Nelson Point
Metcalf, Edward	21	California	Miner	Nelson Point
Metcalf, Samuel	64	Ireland	Miner	Nelson Point
Metcalf, William	57	Ireland	Miner	Nelson Point
Remmington, Edward Willis	34	California	Miner	Nelson Point
Schubert, Gustave Henry	24	California	Photographer	Nelson Point
Schubert, John Henry	57	Germany	Farmer	Nelson Point
Waterworth, James William	30	Iowa	Miner	Nelson Point
Winters, Ganke Rudolph	64	Holland	Miner	Nelson Point
Caine, Francis	65	Ireland	Miner	Nelson Point

1900

Index to Precinct Register. **NELSON POINT**

Berg, Charles	60	Nelson Point	1
Cutler, Elihu	76	do	2
Dow, John Lyman	63	do	3
Drumgold, James Henry	39	do	4
Durrell, Orrin	66	do	5
Eddelbuttle, Johann Conrad	64	do	6
Egbert, Henry William	70	do	7
Houk, David	69	do	8
Jones, Austin "C"	22	do	9
Johnson, Frank Even	29	do	10
Johnson, Albert Allen	21	do	11
Joy, Charles Edward	53	do	12
Lindsey, Edward Galloway	31	do	13
Lawson, Peter	67	do	14
Labadie, Henry Warrén	37	do	15
Metcalf, William Henry	25	do	16
Metcalf Ed	29	do	17
Metcalf, William	63	do	18
Metcalf, Samuel	72	do	19
Mandeville, Frank Charles	44	do	20
Olson, Magnus	73	do	21
Pauly, Nicholas Oscar	72	Mohawk	22
Pauly, Henry Robert	28	Nelson Point	23
Pauly, Oscar Nicholas	23	do	24
Remington, Edward Willis	42	do	25
Russell, Laws on Myrick	64	do	26
Schubert, John Henry	65	do	27
Tucker, William Morris	48	do	28
Thomas, David Philip	34	do	29
Terry, Francis Collins	68	do	32
Winters, Gauke Rudolph	72	do	30
Waterworth, James William	37	do	31

14 NELSON POINT PRECINCT

	Name	Sex	Occupation	Address	Politics
1	Botzbach, Peter Christian	Male	Miner	Quincy	Republic
2	Egbert, Harvey William	Male	Miner	Nelson Point	Democra
3	Fisher, Charles Lincoln	Male	Miner	Quincy	Republic
4	Lindsey, Edwin Galloway	Male	Miner	Nelson Point	Democra
5	Mandeville, Annie Sidona	Female	Housewife	Quincy	Republic
6	Mandeville, Frank Charles	Male	Farmer	Quincy	Republic
7	Sommers, John Benjamin	Male	Miner	Quincy	Republic

Index to Precinct Registers, Plumas County, California, 1912

Nelson Point Precinct

NUMBER	NAME	OCCUPATION	ADDRESS	POLITICS
1	Absolom, James	Miner	Quincy	Rep.
2	Daily, Marcus	Miner	Spring Garden	Dem.
3	Delsol, Camille	Miner	Quincy	D. S.
4	Doyle, Willett W.	Miner	Quincy	Rep.
5	Geraghty, John	Miner	Spring Garden	D. S.
6	Martin, Henry C.	Miner	Quincy	Rep.
7	Maurizzio, Frank	Miner	Quincy	D. S.
8	Maurizzio, Mrs. Louise	Housewife	Quincy	Rep.
9	O'Callaghan, A. R.	Miner	Quincy	Dem.
10	Pauly, Mrs. Katherine	Housewife	Quincy	Dem.
11	Seymore, Andrew Renick	Miner	Quincy	Dem.
12	Schimmer, Charles	Miner	Quincy	Rep.
13	Tower, Bert	Miner	Quincy	Rep.
14	Turner, Ed. J.	Miner	Nelson Point	Dem.

Index to Great Register, Plumas County, California
General Election, November 4, 1924

Nelson Point Precinct

Name	Occupation and Address	Politics
Doyle, W. W.;	Miner, Quincy, Calif.	Dem.
Seymour, Andrew;	Miner, Quincy, Calif.	Dem.

Index to Precinct Registers, Plumas County, California
General Election, November 6, 1928

APPENDIX XII.

COCK FIGHTING RULES.

Printed at the Plumas National Office, Quincy, Cal.

QUINCY COCK FIGHTING RULES.

QUINCY, PLUMAS COUNTY, CAL., 1869.

ARTICLE I.

The Pit shall be at least fifteen feet in the clear, if possible (inside), and sixteen inches high from the floor. There shall be two scores two feet each from the center mark.

ARTICLE II.

The Pitters shall each choose one Judge, who shall choose a Referee. The said Judges shall decide all matters in dispute during the pendency of a fight, but in case of their inability to agree, then it shall be the duty of the Referee to decide, whose decision shall be final.

ARTICLE III.

Chickens shall take their age from the first day of February, and shall be chickens during the following fighting season, to-wit: From the first day of February, 1869, until the first day of May, 1870.

ARTICLE IV.

It shall be deemed foul for either of the Pitters to pit a cock with any soap, grease, pepper, or any other external application upon him, or with what is termed a foul hackel (that is, any of the shining feathers left whole upon his neck or mane). Objections must be made before the cocks are put down.

ARTICLE V.

The Pitters shall let each cock bill each other three or more times; but this shall not be so construed, that the Pitter of a cock has a right to bill with his opponent's cock for the purpose of fatiguing him.

ARTICLE VI.

No person shall be permitted to handle his fowl after he has been delivered into the pit, unless he is hung in his adversary, or in himself, or in the pit; or, unless one or both cocks refuse fight, when either Pitter can count—the count going for the last fighting cock. But the Pitter of the non-fighting cock cannot count while the other cock fights.

ARTICLE VII.

Any cock that may get on his back, the Pitter thereof shall turn him off it; but not take him off the ground he is lying on.

ARTICLE VIII.

Whenever a cock is fast in his adversary, the Pitter of the cock the spurs are fast in, shall draw them out. But the Pitter of the cock has no right to draw his own spurs, except when fast in himself, or in the pit.

ARTICLE IX.

When a count is being made, and the fighting cock hangs the non-fighting cock, the gaffs shall be drawn and the count cease when the cocks are touched, and shall be begun when the cocks are pitted, at the number where the count ceased.

[illegible] once told; twice told, etc. until the law is complete (counting it deliberately and distinctly).

ARTICLE XI.

No Pitter, after the cock has been delivered into the pit, shall be permitted to clean their beaks or eyes, by blowing on them, or squeezing or pressing them against the floor.

ARTICLE XII

It shall be duty of the respective Pitters to deliver their cocks fairly on their feet, [illegible] third and last ten being told, when they shall be delivered on the center score, breast to breast). Any Pitter shoving his cock, or using any other unfair means to make him fight, shall lose the fight.

ARTICLE XIV.

No Pitter shall be permitted to shove, push or handle his adversary's cock, after he has been delivered in the pit, and such conduct shall be considered foul.

ARTICLE XV.

If both cocks fight together, and if then both should refuse to fight, until the law is counted out, it is a drawn fight.

ARTICLE XVI.

If a cock should die before the law is counted out, he shall be deemed the loser. But this does not apply to a runaway cock, who must fight to win.

ARTICLE XVII.

Crowing or raising a hackel is not fight, nor the fighting at the Pitter's hands. A bite at a cock, or feeling of a blind cock, is fight.

ARTICLE XVIII.

A breaking cock is a fighting cock. But a cock breaking away from his adversary is not fight.

ARTICLE XIX.

If any dispute arise [illegible]ween the Pitters on the res[illegible], the cocks are not to be taken [illegible] the pit, or the gaffs taken off [illegible] it is decided by the Judges or Referee.

ARTICLE XX.

Each cock, within two ounces weight of each other, shall be a match (except blinkers, when fighting against two-eyed cocks, when four ounces will be allowed). Four ounces are also also allowed chickens or stags fighting against cocks.

ARTICLE XXI.

Any person fighting a cock heavier than he is represented on the match-list, loses, although he may have won the fight.

ARTICLE X.

When either Pitter shall have counted forty, (clear and distinct,) without the cock fighting that refused fight, they shall then be handled and placed on the outer scores. If the non-fighting cock still refuses fight, the Pitter of the cock that fights shall go on and count ten, six successive times, breasting them on the center score after the second ten has been told, also after the fifth ten has been told. If the cock still refuses fight, it is a fight won to the cock that fought last.

ARTICLE XI.

After forty has been once told, it shall not be necessary to count it again during that fight, but the Pitter who counts the law shall say: Ten

ARTICLE XXII.

In all cases of appeal, fighting ceases until the Judges or Referee give their decision, which shall be final, and strictly to the question before them.

ARTICLE XXIII.

When a bet is made, it cannot be declared off, unless by consent of both parties. All outside bets go according to the main bet.

ARTICLE XXIV.

Any person violating any of the above rules. shall be deemed to have lost the match.

RULES FOR FIGHTING MAINS.

ARTICLE I.

After the parties have agreed upon the light and heavy weights, it is understood that they fight one match at each weight.

ARTICLE II.

Each party shall produce his list of weights, and proceed to make a match list of their respective cocks in the following manner: All the equal weights shall be first matched, when they shall proceed to match others, giving one and two ounces each in such a way as to make the most matches; making in the aggregate equal weights.

ARTICLE III.

The parties shall toss for choice of beginning. The party winning the toss shall have the privelege of commencing at the light or heavy weight, but must fight in proper order, from top to bottom or from bottom to top.

ARTICLE IV.

In case an equal number of matches fail in, the party winning the first equal half of all the matches to be fought, will be deemed the winner of the odd fight.

ARTICLE V.

All fights will be governed by the laws for match fighting.

ARTICLE VI.

Any change in the rules for fighting mains can be expressed by articles agreed upon by both parties.

APPENDIX XIII.

PAULY MARRIAGE LICENSE.

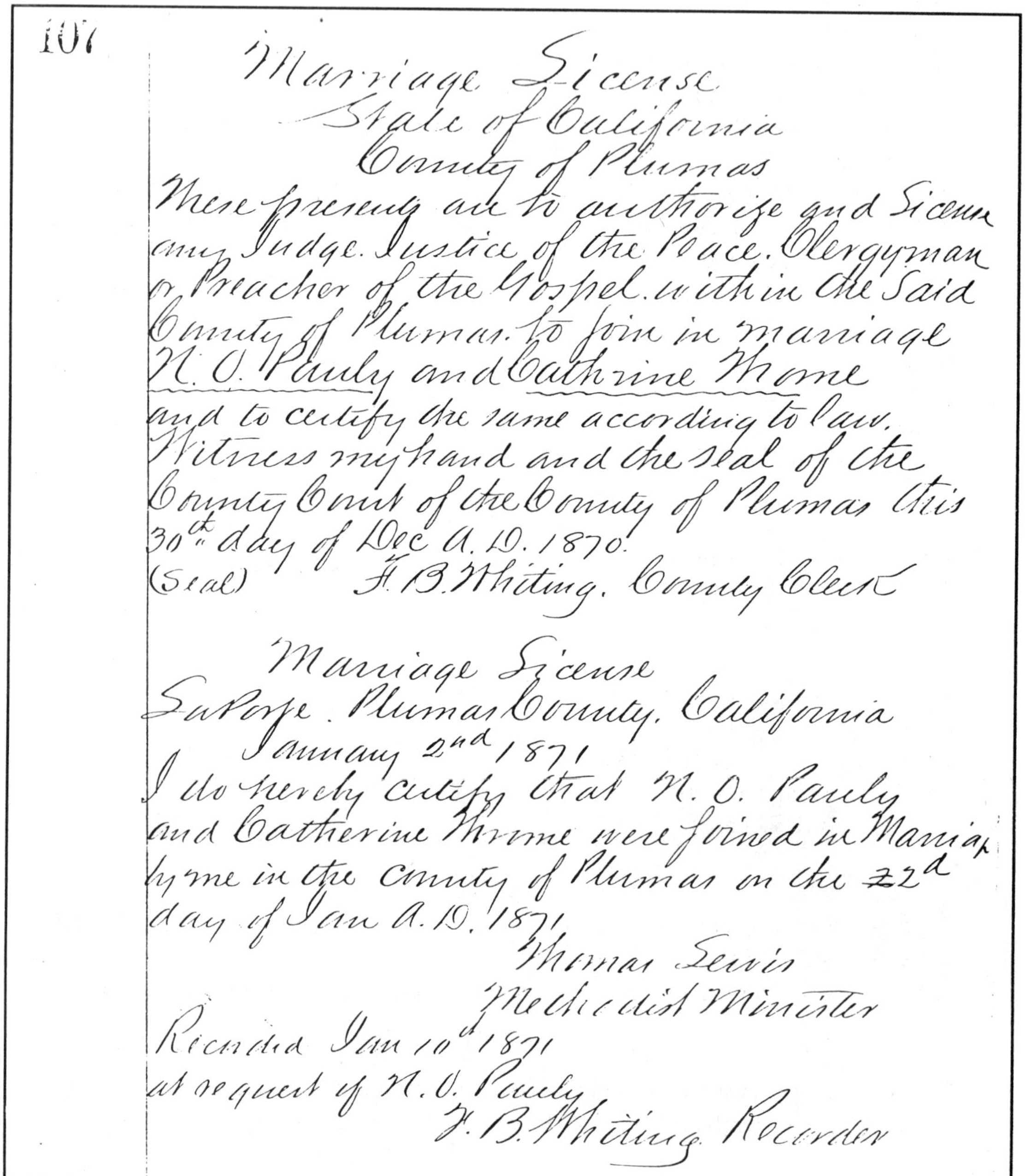

107

Marriage License
State of California
County of Plumas

These presents are to authorize and License any Judge. Justice of the Peace. Clergyman or Preacher of the Gospel. within the Said County of Plumas. to join in marriage N. O. Pauly and Cathrine Throne and to certify the same according to law. Witness my hand and the seal of the County Court of the County of Plumas this 30th day of Dec A. D. 1870.

(Seal) F. B. Whiting. County Clerk

Marriage License
LaPorte. Plumas County. California
January 2nd 1871

I do hereby certify that N. O. Pauly and Catherine Throne were joined in Marriage by me in the county of Plumas on the 2d day of Jan A. D. 1871

Thomas Lewis
Methodist Minister

Recorded Jan 10th 1871
at request of N. O. Pauly
F. B. Whiting. Recorder

BIBLIOGRAPHY.

I. PRIMARY SOURCES.

JOURNALS AND DIARIES.

Clappe, Louise ("Dame Shirley"). *The Shirley Letters.* Santa Barbara: Peregrine Smith, Inc., 1970.

Edwards, William A.Z. *William A.Z. Edwards Diaries and Account Books 1843-1908.* vols. 4-15. Bancroft Library: unpublished.

Hambly, David William. *Hambly Autobiography 1820-1890.* Bancroft Library: unpublished 1885.

Parke, Charles R. *Charles Parke Journal.* The Huntington Library: unpublished.

Parker, William Tell. *Notes By the Way: a journal of the gold- rush days.* The Huntington Library: unpublished.

Haun, James H. *James Haun Diary.* Plumas County Museum Archives: unpublished.

Haun, John. *John Haun Diary.* Plumas County Museum Archives: unpublished.

Kepner, John K. *Diary of John Kline Kepner.* Plumas County Museum Archives: John R. Kirkham, 1985.

Windeler, Adolphus. *The California Gold Rush Diary of a German Sailor.* Berkeley: Howell-North Books, 1969.

NEWSPAPERS.

Alta California. San Francisco, 1851-63.

Feather River Bulletin. Quincy, 1966.

La Porte Union. La Porte, 1868.

Plumas Independent. Quincy, 1902.

Plumas National [Bulletin]. Quincy, 1866-1924.

Plumas Standard. Quincy, 1859-62.

Plumas Argus. Quincy, 1857-58.

Quincy Union. Quincy, 1864-68.

Sacramento Bee. Sacramento, 1947.

REPORTS—PUBLISHED.

California State Agricultural Society. *Transactions: California State Agricultural Society.* Sacramento: 1858.

California State Mining Bureau. *Register of Mines and Minerals.* Sacramento: 1898.

REPORTS—UNPUBLISHED.

Plumas County Census 1860. Plumas County Museum Archives.

Plumas County Merchant Records 1854-59. Plumas County Museum Archives.

MANUSCRIPTS.

Bancroft, Hubert H. *Bancroft Notes on Mines and Mining.* Bancroft Library: uncatalogued.

Bancroft, Hubert H. *Bancroft Scraps vols. I-V, LI:1-4.* Bancroft Library.

Carr & Co. Store Ledger 1862-67. Plumas County Museum Archives.

Huskinson. *Huskinson Scrapbook.* Plumas County Museum Archives.

Plumas County Museum Archives. Thompson file.

Plumas County Museum Archives. Pauly file.

Plumas County Museum Archives. Willow Ranch file.

Plumas County Museum Archives. Cockfighting file.

Plumas County Museum Archives. School file.

Plumas County Museum Archives. Egbert file.

Plumas County Museum Archives. Nelson Point file.

INTERVIEWS.

Cotter, Robert. Cotter residence, Quincy, California. Interview, 25 July 1987.

Haddick, Ruth. Haddick residence, Cromberg, California. Interview, 21 July 1987.

Lawson, Scott. Plumas County Museum Archives, Quincy, California. Interview, 5 August 1991.

Post, Leroy. Post residence, La Porte, California. Interview, 24 July 1987.

II. SECONDARY SOURCES.

BOOKS.

Adams, Ramon F. *Western Words: A Dictionary of the American West.* Norman: University of Oklahoma Press, 1981.

Aki, Nancy. *Principle Chinatowns, Mining Camps in NE California.* Plumas County Museum Archives: unpublished.

Community United Methodist Church. *A Century of Service: 1875- 1975.* Quincy: 1975.

Cotton, Aylett R. *Across the Plains to California in 1849 And After: An Autobiography.* Bancroft Library: unpublished 1906.

Delano, Alonzo. *Alonzo Delano's California Correspondence.* Sacramento: Sacramento Book Club, 1952.

Delano, Alonzo. *Across the Plains and Among the Diggings.* New York: Wilson-Erickson, Inc., 1936.

Dunn, Mary. *The Chinese in Early California.* Plumas County Museum Archives: unpublished.

Fariss and Smith. *History of Plumas, Lassen & Sierra Counties.*

San Francisco: Fariss and Smith, 1882.

Ferguson, Charles D. *The Experience of a Forty-niner During a Third of a Century in the Gold Fields.* Chico: M.A. Carson, 1924.

Fricksted, Walter N. *A Century of California Post Offices.* (Butte County Library.)

Gould, Helen W. *La Porte Scrapbook.* La Porte: Helen W. Gould, 1972.

Gudde, Erwin G. *California Gold Camps.* Berkeley: University of California Press, 1975.

Hittell, Theodore H. *History of California.* 4 vols. San Francisco: 1897.

Holliday, J.S. *The World Rushed In.* New York: Simon and Schuster, 1981.

Lecouvreur, Frank. *From East Prussia to the Golden Gate.* New York: Angelina Book Concern, 1906.

Power, Robert H. *Pioneer Skiing in California.* Johnsville: Plumas Eureka State Park Association, 1980.

Webster, Kimball. *The Goldseekers of '49.* Manchester, N.H.: Standard Book Co., 1917.

ARTICLES IN JOURNALS.

Plumas County Historical Society. *Plumas Memories* 35 (June, 1970).

Riley, Glenda. "The Spectator of a Savage: Rumors and Alarmism on the Overland Trail." *Western Historical Quarterly 15 (October, 1984): 443.*

Wunder, John W. "Chinese in Trouble: Criminal Law and Race on the Trans-Mississippi West Frontier." *Western Historical Quarterly* 17, No. 1 (January, 1986): 38.

INDEX.

ORDER FORM

Nelson Point

Portrait of a Northern Gold Rush Town

$19.95 Postpaid

Please send me ____copies of *Nelson Point* @ $19.95 each postpaid.
California residents add 7.75%, or $1.55 per copy, for state sales tax.
Enclosed is my check for $______, payable to Pacific Sunset Publishing.

NAME ____

COMPANY NAME ______________________________________

ADDRESS ____

CITY ______________________ STATE __________ ZIP __________

DAYTIME TELEPHONE _________________________________

Send check or money order to:
Pacific Sunset Publishing
RR 2 Box 668-A
Redlands, California 92373

Discounts for quantity orders. Please inquire.
If this is a library book, please photocopy this page.

Thank you!